EPA/600/R-13/078 | December 2013 | www.epa.gov/research

I0502849

Evaluation of Opportunities to Improve Structural Inspection Capabilities for Water Mains: Large Diameter Cast Iron Pipe

Office of Research and Development

EPA/600/R-13/078
December 2013

Evaluation of Opportunities to Improve Structural Inspection Capabilities for Water Mains: Large Diameter Cast Iron Pipe

by

John Matthews, Ph.D., Bruce Nestleroth, Ph.D. and Wendy Condit, P.E.
Battelle Memorial Institute

and

James Thomson, C.Eng.
Independent Consultant

EPA Contract No. EP-C-05-057
Task Order No. 62

Michael D. Royer
Task Order Manager

Water Supply and Water Resources Division
National Risk Management Research Laboratory
U.S. Environmental Protection Agency
2890 Woodbridge Avenue (MS-104)
Edison, NJ 08837

National Risk Management Research Laboratory
Office of Research and Development
U.S. Environmental Protection Agency
Cincinnati, Ohio 45268

DISCLAIMER

ABSTRACT

The U.S. EPA and other organizations have projected that a large portion of the United States' aging water conveyance infrastructure will reach the end of its service life in the next several decades. EPA has identified asset management as a critical factor in efficiently addressing this projected surge in water conveyance infrastructure renewal. An important tool in the asset manager's toolbox is cost-effective structural inspection, since it provides data to help support optimized capital, operations, and maintenance planning. However, there are many gaps in structural inspection capability and affordability, and many options for addressing those gaps.

Recognizing the importance of structural inspection, and its shortcomings, as well as the many potential options for its improvement, U.S. EPA's Office of Research and Development identified evaluation and improvement of structural inspection technologies as an important component of its aging water infrastructure research program. Selection of the most promising structural inspection technologies to evaluate and/or improve is challenging due to the number of factors to be considered. This project provides a protocol for screening innovative structural inspection concepts and technologies. The protocol is focused on a single scenario -- large diameter, cast iron water mains -- because (a) it substantially reduces the complexity of the decision protocol, (b) high consequence failures of this type have already occurred and remain of concern for older cities; and (c) research by others has provided new insights into the causes of these failures that enabled important pre-failure indicators to be identified and quantified.

The initial target audience is EPA's aging water infrastructure research planning process. The protocol is expected to be used as a guide for periodic reviews of the prospects of emerging structural inspection technologies for large diameter cast iron water mains. Also, this protocol can potentially be utilized by other organizations or individuals who are considering supporting or conducting water or wastewater pipe structural inspection technology research. The protocol can potentially be modified to address other high-interest pipe scenarios, such as large diameter ductile iron, pre-stressed concrete cylinder pipe (PCCP), asbestos cement, and steel.

The protocol contains three levels and is used to evaluate eight technologies (four existing and four emerging) to determine whether the protocol is implementable and produces reasonable results. The report provides a brief overview of the potential failure modes, mechanism, and distress indicators for high risk cast iron water mains. This report also briefly discusses structural inspection technologies and the key stakeholders involved in structural inspection technology research and development.
The protocol guides information collection for proposed technology developments including: the potential for inspection of water mains, the ability to detect specific anomalies or abnormal operating conditions, the cost of the technology development, and the potential cost of utilization by water companies. The application of these protocols was successful in demonstrating the potential of eight technologies, four of which showed the potential for further development for detecting critical distress indicators in a relatively cost-effective and reasonably implementable manner. The other four were considered to be inappropriate for further development for use on large diameter, cast iron water mains.

ACKNOWLEDGMENTS

The technical direction and coordination for this project was provided by Michael Royer of EPA's Urban Watershed Management Branch. The project team would like to acknowledge the technical input of several contributors to this report including Balvant Rajani and Yehuda Kleiner from the National Research Council (NRC) of Canada; Frank Blaha from the Water Research Foundation (WaterRF); and Walter Graf from the Water Environment Research Foundation (WERF). The authors would like to thank the stakeholder group members (Frank Blaha of WaterRF, Walter Graf of WERF, and Yehuda Kleiner of NRC Canada) for providing written comments.

This report includes an updated review of research on advances in structural inspection technologies for cast iron water mains based on unpublished, comprehensive literature reviews conducted initially under EPA Contract No. CWX089 entitled, *Report on Evaluation of Opportunities to Improve Structural Integrity Monitoring (SIM) Capabilities for Water Mains through Federal Research and Technology Transfer*, EPA STREAMS Contract No. EP-C-05-057, Task Order (TO) 62 sub-tasks entitled, *Evaluation of Opportunities to Improve Structural Integrity Monitoring (SIM) Capabilities for Water Mains*, and *Critical Review of Recently Completed and Ongoing Water Main Condition Assessment Products and Research*.

EXECUTIVE SUMMARY

Introduction

The U.S. EPA and other organizations have projected that a large portion of the United States' aging water conveyance infrastructure will reach the end of its service life in the next several decades. EPA has identified asset management as a critical factor in efficiently addressing this projected surge in water conveyance infrastructure renewal. An important tool in the asset manager's toolbox is cost-effective structural inspection, since it provides data to help support optimized capital, operations, and maintenance planning. However, there are many gaps in structural inspection capability and affordability, and many options for addressing those gaps.

Recognizing the importance of structural inspection, and its shortcomings, as well as the many potential options for its improvement, U.S. EPA's Office of Research and Development identified evaluation and improvement of structural inspection technologies as an important component of its aging water infrastructure research program. Selection of the most promising structural inspection technologies to evaluate and/or improve is challenging due to the number of factors to be considered. This project provides a protocol for screening innovative structural inspection concepts and technologies. The protocol is focused on a single scenario -- large diameter, cast iron water mains -- because (a) it substantially reduces the complexity of the decision protocol, (b) high consequence failures of this type have already occurred and remain of concern for older cities; and (c) research by others has provided new insights into the causes of these failures that enabled important pre-failure indicators to be identified and quantified.

The initial target audience is EPA's aging water infrastructure research planning process. The protocol is expected to be used as a guide for periodic reviews of the prospects of emerging structural inspection technologies for large diameter cast iron water mains. Also, this protocol can potentially be utilized by other organizations or individuals who are considering supporting or conducting water or wastewater pipe structural inspection technology research. The protocol can potentially be modified to address other high-interest pipe scenarios, such as large diameter ductile iron, pre-stressed concrete cylinder pipe (PCCP), asbestos cement, and steel.

Characterization of Potential High Risk, Cast Iron Water Mains

The structural deterioration and failure pattern of high-risk, large diameter cast iron pipes is complex due to factors such as the heterogeneous nature of cast iron, variability of handling and installation, and differing soil properties along the line. Despite these complexities, structural inspection can be an important component in estimating the current and future condition of cast iron pipe. Some failure mechanisms have potentially reliable measurable distress indicators; therefore, it is reasonable to expect that if monitored, these indicators could help determine if failure may be imminent or if an asset can operate longer before failure.

The most common failure modes and mechanisms for large diameter cast iron pipe are: longitudinal fractures, circular fractures, mixed fractures, bell splitting, and corrosion. Longitudinal cracking is more common in large diameter pipes and can take various forms such as vertical cracks and slanted cracks across the pipe wall. Circular cracking is the most common failure mode for small diameter pipes, although there are cases recorded in large diameters. Mixed fractures can be either tensile hoop failures in combination with bending or shattering due to the annealing process. Bell splitting is due to longitudinal cracks at the bell end or bell shards. Corrosion in the form of pitting and/or graphitization is a common but not exclusive factor in most pipe failures.

Factors that could potentially contribute to failure include physical (e.g., pipe age, thickness, and vintage); environmental (e.g., pipe bedding, soil type, and climate); and operational factors (e.g., internal water pressure, transient pressure, and leakage). Measurable distress indicators that are the result of these factors include: external coating defects, pipe barrel and bell wall thickness, graphitization, and cracks; internal lining spalling; tuberculation; change in joint alignment; and joint displacement. Inferential indicators (e.g., pipe vintage, pressure variations, and pipe location) can point to the potential existence of a pipe deterioration mechanism, but they do not provide direct evidence of structural distress.

Structural Inspection Components and Systems

To successfully monitor structural condition, a combination of screening, monitoring, and condition assessment techniques needs to be used. External condition assessment tools, which provide detailed condition information for selected locations along the pipe, include corrosion pit depth measurements; ultrasonic tools for measuring remaining pipe wall thickness; magnetic flux leakage (MFL) technology for detection of graphitization and cracks on the exterior of pipe walls; and broadband electromagnetic technology for detecting wall loss.

Inline inspection technologies include closed circuit television (CCTV) visual tools to inspect for cracks; remote field technology tools to detect loss of wall thickness; and acoustic leak detection tools. Issues that must be overcome for wide-spread use of inline inspection technologies for water mains include the lack of launching and receiving facilities on existing water mains and the expense of conducting the inspections. Other leak detection technologies can also be applied externally and do not require internal access to the main.

Key stakeholders involved in structural inspection technology research and development include: federal agencies (i.e., U.S. Environmental Protection Agency [EPA], U.S. Department of Transportation[DOT], U.S. Department of Energy [DOE], U.S. Department of Commerce/National Institute for Standards and Technology [NIST], National Science Foundation [NSF], etc.); non-profit organizations (e.g., Water Research Foundation [WaterRF], Water Environment Research Foundation [WERF], etc.) and private technology developers. These agencies and organizations would be the users of the protocol developed in this report, which would serve as a guide for the screening and identification of emerging technologies that could be evaluated for their suitability to large diameter cast iron water mains.

Protocol Development

The screening protocol is a three step process designed to assist EPA or other research funding organizations to strategically evaluate the feasibility of emerging structural inspection technologies for use on large diameter cast iron mains. The first screening protocol collects the data needed to enable a user to determine if an inspection technology can be practically implemented on a large diameter cast iron water main. After determining the intended capabilities of the tool, the user is led through a series of flowcharts to determine how the tool is implemented (i.e., internal, external, from above ground, or the air); what the requirements are for implementation; and the technology is given an implementation grade (i.e., easy, moderate, or difficult) based on the user's answers.

The second screening protocol collects the data to enable a user to determine the degradation condition or conditions that an inspection technology can detect and determines if the technology locates the key distress indicators for large diameter cast iron water mains. First, the protocol is used to determine what types of defects the tool is intended to detect (i.e., degradation conditions such as corrosion and leaks; condition that could lead to failure such as pipe angle at the joints; etc.) and to what degree the tool can detect them. Then the technology is given a cost-to-implement grade based on the current costs of the system and the needed cost for further development. The third and final screening protocol determines how a structural inspection technology compares to existing technologies and whether it is recommended for further development. Input from the first protocol about the types and degrees of defects the tool can detect are used, along with implementation and cost grades to determine if the proposed technology has the potential for further development.

Application of Protocols

To validate and calibrate the protocol, eight example applications using existing and emerging technologies were conducted. Two remote field eddy current based technologies, which are used to detect wall corrosion internally, were evaluated with one showing the potential for further development, while the other approach was considered inappropriate for further development due to the cost to implement. The two internal leak detection technologies both showed potential for further development, and currently both are being designed for use on large diameter cast iron mains. Two technologies designed for aboveground use to measure wall corrosion were also evaluated with one showing the potential for further development, while the second was considered inapplicable due its need for continuous electrical conductivity. The final two technologies were MFL technologies used to detect wall corrosion internally, but each was considered inappropriate for further development due to their difficult implementation and high costs.

Conclusions and Recommendations

The screening protocols developed for this report can help to guide EPA or other funding agencies in evaluating the potential applicability of proposed structural inspection technologies for use with high-risk, large diameter cast iron water mains, which can be very costly when they fail and when they are replaced. The protocols collect information for proposed technology developments including: the potential for inspection of water mains, the ability to detect specific anomalies or abnormal operating conditions, the cost of the technology development, and the potential cost of utilization by water companies. The application of these protocols was successful in demonstrating the potential of eight technologies, four of which showed the potential for further development for detecting critical distress indicators in a relatively cost-effective and reasonably implementable manner. The other four were considered to be inappropriate for further development for use on large diameter cast iron water mains.

It is recommended that EPA or other funding agencies interested in supporting the evaluation and improvement of structural inspection technologies use these protocols as a screening measure to determine if a proposed technology is applicable to large diameter cast iron water mains, capable of detecting key distress indicators, and implementable at a reasonable cost. This methodology was developed for large diameter cast iron mains as an example and can be expanded to small diameter mains and other pipe types.

It is also recommended that screening protocols be developed for other potentially high-risk mains such as large diameter ductile iron, prestressed concrete cylinder pipe (PCCP), asbestos cement, and steel.

TABLE OF CONTENTS

FIGURES

TABLES

ABBREVIATIONS AND ACRONYMS

AC	alternating current
AWWA	American Water Works Association
AWWARF	American Water Works Association Research Foundation
BEM	broadband electromagnetic
BOEMRE	Bureau of Ocean Energy Management, Regulation, and Enforcement
CCTV	closed circuit television
CEIT	Center for Environmental Industry & Technology
CERL	Construction Engineering Research Laboratory
CI	cast iron
CIPRA	Cast Iron Pipe Research Association
CWA	Clean Water Act
DHS	U.S. Department of Homeland Security
DIPRA	Ductile Iron Pipe Research Association
DOC	U.S. Department of Commerce
DOD	U.S. Department of Defense
DOE	U.S. Department of Energy
DOI	U.S. Department of the Interior
DOT	U.S. Department of Transportation
EPA	U.S. Environmental Protection Agency
ESTCP	Environmental Security Technology Certification Program
ETV	Environmental Technology Verification
FRS	flexible rod sensor
FSA	free swimming acoustic
GTI	Gas Technology Institute
IRC	Institution for Research in Construction
ITSC	International Science and Technology Center
LaRC	Langley Research Center
MFFV	multi-frequency field variation
MFL	magnetic flux leakage
MIC	microbiologically influenced corrosion
MTM	magnetic tomography
NASA	National Aeronautics and Space Administration
NCER	National Center for Environmental Research
NDE	nondestructive evaluation
NETL	National Energy Technology Laboratory
NIST	National Institute of Standards and Technology
NRC	National Research Council
NREC	National Robotics Engineering Center
NRMRL	National Risk Management Research Laboratory

NSF	National Science Foundation
NTIAC	Nondestructive Testing, Information, and Analysis Center
OP	operating pressure
OPS	Office of Pipeline Safety
ORD	Office of Research and Development
PRCI	Pipeline Research Council International
R&D	research and development
RDT&E	Research, Development, Test & Evaluation
RFEC	remote filed eddy current
RFT	remote field technology
S&T	Science and Technology Directorate
SAM	Strategic Asset Management
SBIR	Small Business Innovation Research
SDWA	Safe Drinking Water Act
SERDP	Strategic Environmental Research and Development Program
SRB	sulfate reducing bacteria
STAR	Science to Achieve Results
TA&R	Technology Assessment & Research
TIP	Technology Innovation Program
TO	Task Order
USACE	U.S. Army Corps of Engineers
UKWIR	UK Water Industry Research
WaterRF	Water Research Foundation
WERF	Water Environment Research Foundation
WSWRD	Water Supply and Water Resources Division

1.0: INTRODUCTION

1.1   Project Background

Cost-effective structural inspection can be an important component of effective condition assessment and asset management of water conveyance infrastructure. Structural inspection involves collecting data about meta-stable and/or transient indicators of the condition of the pipe. The data are used as inputs for estimating the current and future condition of the pipe.

Cost-effective structural inspection can provide value to utilities in three primary ways. First, it can help the utility prevent catastrophic failures in their deteriorating water mains, which they cannot afford to replace at present. Secondly, it can help the utility reduce the amount perceived to need replacement, which may enable them to replace only the pipes that are structurally deteriorated to the point that their probability of failure is unacceptable. Finally, it may help the utility reduce the rate of deterioration of its aging water mains. This could enable the utility to more promptly identify pipes that are deteriorating at an accelerated rate, which could lead the utility to mitigate the conditions causing accelerated deterioration with action (e.g., leak repair, retrofit cathodic protection, and/or spot rehabilitation).

These benefits provided by structural inspection technologies are important for utilities that must strategically select water mains for replacement since they cannot afford to replace their entire aging infrastructure. No references are readily available outlining how much utilities would be willing to pay for condition assessment versus replacing costs, but rehabilitation and maintenance activities are typically cheaper than replacement and can extend the asset life for years before replacement is needed (Baird, 2010).

These technologies can be improved by identifying failure modes and indicators; improving technology performance (e.g., efficiency of detecting critical flaws; better spatial, temporal, and indicator coverage); and reducing cost (e.g., mobilization/demobilization; pipe preparation/access; data collection; data analysis; speed; and share cost [i.e., use same data or platform or data transmission system]). Scientific and engineering research is being conducted to develop and improve these technologies and to accelerate commercial implementation; portions of the development work are funded in part by government and industry associations. Unfortunately, some developments are unsuccessful because they are not feasible for water pipelines, do not properly address the structural integrity issue, or are not as cost effective as competing structural inspection methods.

This report presents a protocol to assist research funding organizations, such as the U.S. Environmental Protection Agency (EPA), in strategically evaluating the feasibility of emerging structural inspection technologies for large diameter cast iron (CI) mains. This technology evaluation protocol could be used as a guide for periodic expert panel reviews of the prospects for proposed innovations to pipe inspection technologies. There is additional benefit from operating the evaluation protocol on potential technology transfer opportunities from other industries, and innovative technologies from small businesses and university research and development.

1.2   Project Objectives

The objective of this work was to develop a targeted, sound, and easy to use screening protocol for evaluation of structural inspection technologies that are used to provide data that can support estimates of current and future structural condition of water mains. These estimates can be used to help optimize decisions about inspection, rehabilitation, and replacement of water mains. The value of optimal renewal decision making arises from (1) safely utilizing installed infrastructure to its full life, (2) reduction of

1

main break failures and their adverse health, safety, environmental, and economic effects, and (3) prompt recognition and correction of significant leakage or deterioration.

The screening protocol developed focuses on large diameter, cast iron water mains. The screening protocol evaluated the feasibility for further development of structural inspection technologies that could be used to cost effectively prevent catastrophic failures, reduce the amount of pipe that needs to be replaced, and/or reduce the rate of deterioration. The protocol contains three levels and was used to evaluate eight technologies consisting of existing and emerging technologies to see whether the protocol is implementable and produces reasonable results.

This report was developed based on EPA Quality Assurance Project Plan (QAPP) requirements set out in EPA (2001). The quality of the secondary data reported in this document was not independently evaluated by EPA and Battelle.

1.3   Organization of the Report

The first section of the report, Section 2.0, provides consolidated information on potential failure modes and damage indicators for large diameter cast iron mains that could guide the development of new inspection technologies. Section 3.0 presents a brief overview of the various structural inspection technology components for cast iron pipe. The details of the specific technologies are available in the references. Section 4.0 presents the three part protocol. The first protocol is a basic screening protocol to answer whether the technology is feasible for water pipelines. The result of the evaluation is pass/fail. The second screening protocol produces ratings on how well the structural inspection technology will perform in water pipelines and how well it locates defects that will grow to failure and potential conditions that are associated with failures (indicators). The third screening protocol determines how the new structural inspection technology compares to existing technologies. Section 5.0 discusses the testing of the protocol on eight technologies that are commercial, under development or available from other industries. Section 6.0 summarizes key findings and provides recommendations for future research.

2.0: CHARACTERIZATION OF POTENTIAL HIGH RISK, CAST IRON WATER MAINS

2.1   Background

The first cast iron water pipes in the U.S. were installed in Philadelphia in 1804 and many cast iron water pipes in the U.S. have been in continuous operation for over 100 years. An American Water Works Association (AWWA) survey of 337 water utilities determined that about 35% of the U.S. water pipe network was laid with cast iron pipes made up of approximately 18% unlined and 17% lined (AWWA, 2004;U.S. EPA, 2009). Out of approximately 900,000 miles of water pipe in the U.S., it is estimated that 315,000 miles is cast iron.

The structural deterioration and subsequent failure of cast iron water mains is a complex process involving numerous factors both physical and dynamic. Particularly for large diameter cast iron pipes, the pattern of failure (discussed in Section 2.3) may be complex due to factors such as the heterogeneous nature of cast iron, variability of handling and installation, and differing soil properties along the line. A number of physical and dynamic factors such as mechanical strength, loadings, and corrosion rates cannot be precisely defined for each situation. These uncertainties and combinations of factors that create failure make any condition assessment of the remaining life complex. Research has suggested that many failures occur as a series of multiple events rather than a single event (Makar, 2000).

Despite these complexities, effective structural inspection can be an important component in estimating the current and future condition of water mains. Some large diameter cast iron failure mechanisms have potentially critical and reliable measurable distress indicators (e.g., corrosion, graphitization, cracks, leakage, and angled pipe joints) and inferential indicators (e.g., pipe vintage, pressure variations, pipe location, and soil issues), although the critical values that must be measured for each indicator may not be known. Therefore, it is reasonable to expect that reliable condition indicators, if monitored and measured accurately, can help determine if failure is imminent or if an asset can operate for longer before failure.

It is worth noting that a particular inspection technology is only going to be useful for the indicators that it is designed to detect and the failure modes associated with that indicator. It may take a combination of technologies to obtain the desired level of warning. This report focuses on large diameter cast iron mains, defined as 16 in. diameter or greater. These are nearly exclusively transmission mains; fire hydrants and connections are not normally found on such pipes.

2.2   Overview of Cast Iron Pipe

In the early nineteenth century, the first cast iron pipes in the U.S. were imported from England and Scotland. In 1819, the City of Philadelphia installed a 16 in. diameter water main manufactured at Weymouth, New Jersey. It was not until around 1830 that local production became more widely established. Casting of pipes, boilers, and other items was undertaken in many local foundries around the U.S. with variations in quality. Four techniques were employed in the casting of iron pipe (Stanton Ironworks, 1936; CIPRA, 1927) and are briefly described below.

2.2.1   Horizontally Cast Pipes Using Sand Molds. Prior to 1850, the pipes were cast horizontally using an inner core and outer mold. The outer mold was in two halves and formed from moist green sand to form the outer pipe diameter. The inner core forming the internal bore was formed from baked sand reinforced with iron rods. The space between molds was filled with molten iron. The length of pipes was limited to a few feet because of the sagging of the inner core. It was difficult to place the cores concentrically and a tendency to float led to non-uniform wall thickness. Another problem with horizontal casting was the tendency for scum and air bubbles to float to the top of the pipe, which created an area of weakness.

2.2.2   Vertically Cast Pipes Using Sand Molds. From around the 1850s, vertical casting became the method most commonly used for pipe production allowing longer pipe sections to be cast. Sections up to 16 ft for diameters less than 12 in. and 8 to 12 ft for larger diameters could be cast, although by 1927 lengths up to 16 ft were produced.

These pipes were cast in vertical pits. The outer mold was formed from damp sand rammed around an inner metal casing. The casing was withdrawn and the sand baked to form the outer mold. The inner core was formed from sand and clay packed into an inner cylinder and also baked. The outer and inner cores were assembled vertically in the pit and molten iron was poured into the annular space. Originally the pipe bells were formed at the top of the pit, but from around 1914, casting the bell at the bottom was introduced and over a period of time became standard practice. This obviated the weakening of bells due to accumulation of scum and air bubbles in the top of the mold.

A characteristic of cast pipes is a lower fracture toughness and mechanical strength that arises from larger graphite flakes than spun cast iron pipes, which act as crack initiator sites. This is particularly a problem in larger pipes where the slow cooling promoted flake growth.

2.2.3   Horizontally Spun Cast Pipe Using Metal Molds. The de Lavaud technique developed in Brazil in 1915 was licensed to companies in the U.S. in the early 1920s. The method was a great improvement in that casting defects were greatly reduced and pipe with consistent uniform wall thickness was produced. The system was based on pouring molten iron into a metal mold and spinning at high speed to create uniform wall thickness. The rotating mold was dipped into cold water to cool it. This caused the exterior surface of the pipe to be hardened by direct crystallization, which was then softened by heat annealing treatment. Annealing was a difficult process for larger diameters as they had to be rolled onto skids when still at temperatures of 900°C, which could lead to hairline cracking.

2.2.4   Horizontally Spun Cast Pipe Using Sand Molds. This process, introduced in 1925, was based on using dry sand molds which obviated the need for heat annealing as the porous sand allowed for ventilation and cooling. Both spun systems allowed pipe lengths up to 20 ft to be produced. From around the 1850s, pipes were dipped in a Dr. Angus Smith coal-tar oil solution while hot to coat the internal and external surfaces. They were also pressure tested and hammer tapped to detect cracks. Cement lining was first applied to cast iron mains in Charleston in 1921 (CIPRA, 1927).

It cannot be assumed that cast iron pipes will have the mechanical strength as required by the standards in use at the time of their installation. Tests (Kleiner and Rajani, 2000) have shown strengths ranging from 33 MPa to 231 MPa (6,600 to 34,000 psi). There are likely to be wide variations even from the same pipe due to the changing distribution of graphite flakes and casting flaws.

2.2.5   Joints in Cast Iron Pipes. The bell and spigot joint system has found general use in all forms of cast iron pipes. The joint was first made by caulking yarn or hemp into the space between the bell and spigot and then pouring molten lead and caulking using hammers and caulking tools into the remaining space. An alternative jointing material was "leadite" which was introduced initially around 1900 (Rajani and Kleiner, In Press 2013). Leadite is a mixture of iron, sulfur, slag, and salt, which is heated and becomes a vitreous mass when cooled. Leadite had advantages in that it melts at 121°C compared to lead at 322°C and does not have to be caulked. The use of a compressible flexible gasket was introduced in the late 1950s for a push-on joint, with rubber being one of the widely used materials.

2.3   Failure Modes and Mechanisms

Failure in pipes is defined here as a condition caused by collapse, break, or bending, so that a structure or structural element can no longer fulfill its purpose. Other definitions include failures where relatively small amounts of water are lost from defective joints. Failure occurs when the pipe is weakened by corrosion or other defects to an extent when it can no longer resist the imposed stresses. Failure modes, as described in this section based on Rajani and Kleiner (In Press 2013), are the manner in which a cast iron pipe fails and the mechanisms that cause failure.

Smaller diameter pipes generally have smaller moments of inertia making them more susceptible to longitudinal bending failures. Larger diameter pipes have greater moments of inertia which creates a tendency to longitudinal cracking and shearing at the bell. For pipes less than 16-in. in diameter, both the length of pipe and the break frequency are much greater than pipes with larger diameters. Although a number of utilities record failures, there is a dearth of detailed records on the frequency and mode of large diameter pipe failures.

The National Research Council (NRC) of Canada has undertaken detailed investigations of failures in the U.S., UK, and Canada (Rajani and Kleiner, In Press 2013; Makar et al., 2001; Makar, 2001) as has the University of Toronto (Seica et al., 2002). These reports include useful recommendations on the examination of failures of cast iron mains. Reference to these findings is included in the following discussion of failure modes and mechanisms. Tables 2-1 and 2-2 outline the failure modes and mechanisms (Rajani and Kleiner, In Press 2013).

Table 2-1. Longitudinal and Circumferential Breakage Patterns in Large Diameter Mains

	Breakage Pattern	Principal Stress	Location	Possible Cause/Comment
Longitudinal split fracture	*Vertical crack(s) across pipe wall thickness.* • Pull apart – crack with no movement.	Tensile (hoop)		Previously initiated crack that eventually propagates along pipe length. If crack remains open, significant residual stresses were present.
	Slanted crack(s) across pipe wall thickness. • Dip slip – vertical movement. • Reverse slip – vertical movement.	Tensile (hoop) Compression	At spring line (3 or 9 o'clock) if backfill (dense soil) provides good support. At any of 12, 3, 6, 9 o'clock positions if backfill (loose) provide poor support. Crack position would also depend on location of defects or inclusions (weakest link).	Crack initiates through presence of defect, void or inclusion that eventually propagates longitudinally. Direction of the dip helps establish if failure movement is dip or reverse slip.
Circular fracture	*Crack(s) at the invert, spring line or crown.* *Crack(s) across the whole pipe circumference.*	Tensile (longitudinal) Tensile (longitudinal)	 Crack initiates at invert if poor bedding is present or soil support is lost and propagates towards the spring line in circumferential direction.	Not usually observed in large diameter pipes.

Rajani and Kleiner, In Press 2013.

5

Table 2-2. Mixed Fracture and Bell Split Breakage Patterns in Large Diameter Mains

	Breakage Pattern	Principal Stress	Location	Possible Cause/Comment
Mixed fracture	*Principal crack is longitudinal with arc(s) formed at each end of crack; occasionally transforms into a spiral crack.*	Tensile (hoop) + bending (arc end)		Initiates as indicated for longitudinal split and propagates in direction where soil offers little or no support.
			Crack initiates at midpoint of crack and propagates on either or both directions. Subsequently, earth loads come into action causing a bending (flap failure).	External restraint or boundary conditions play a major role on how the crack changes direction to dissipate energy.
	Shattered into multiple pieces of broken pipe.	Tensile (hoop) + residual		Multiple cracks with fallout of pieces occur in pipes with annealing treatment (spun cast with metal molds).
Bell split	*Longitudinal crack at bell end.*	Tensile (hoop)		Crack introduced by wedge action when lead caulked (hammered) into place or excessive hammering action. Subsequently propagates with time with stress fluctuation caused by variations in pressure.
	Bell shard.	Tensile (hoop + flexural)		Reaction caused by spigot on the bell as result of settlement of barrel part of pipe.

Rajani and Kleiner, In Press 2013.

2.3.1 Longitudinal Split Fracture. Longitudinal cracking is typically more common in large diameter pipes. The failure mechanism shown in Figure 2-1 (Rajani and Kleiner, In Press 2013) can take various forms:

- A vertical crack across the pipe wall thickness due to tensile hoop stress
- A slanted crack across pipe wall thickness which takes two forms:
  - A dip slip: vertical movement due to tensile hoop stress
  - Reverse slip: vertical movement due to compression

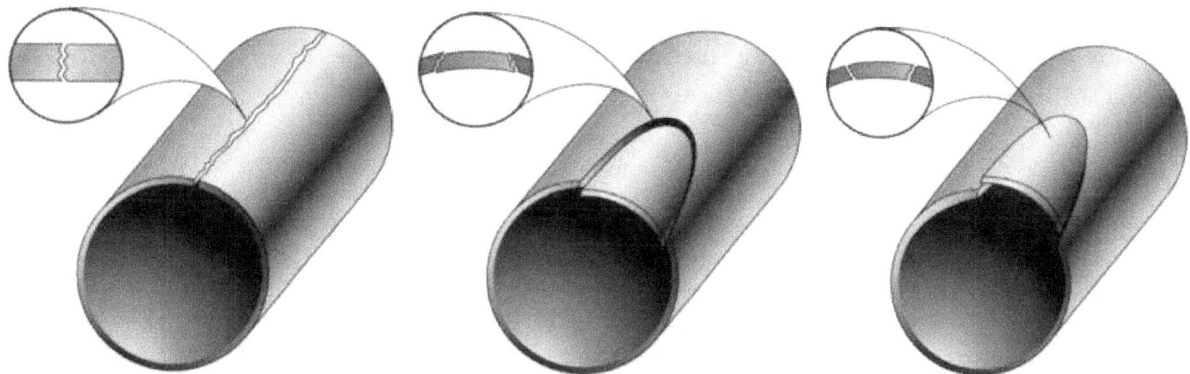

Figure 2-1. Longitudinal Fracture: Vertical Crack (left), Dip Slip (center), and Reverse Slip (right)

Longitudinal cracking is initiated by a crack or a defect in the pipe manufacture. This failure mode can be caused by internal water pressure, external loading which can create bending and crushing forces on the pipe, or compressive forces acting along the pipe particularly where the backfill or bedding support is suspect. Any of these loadings could result in a longitudinal crack. Once the crack has been initiated, it may travel the length of the pipe. Cracks can form on opposite sides of the pipe, resulting in a section of the top of the pipe being removed.

2.3.2   Circular Fracture. Circumferential cracking is the most common failure mode for small diameter (< 380 mm [15 in.] diameter) grey cast iron pipes (Table 2-3) and can be located at the invert, springline, crown, or across the whole pipe (Marshall, 2001). The principal failure mechanism is due to longitudinal tensile stress.

Table 2-3. Percentage of Failures by Mode for Iron Pipe (< 15 in.)

Circumferential	Longitudinal	Hole	Joint
66.4%	13.3%	16.1%	4.2%

Some studies suggest that the majority of failures in all grey cast iron are due to circumferential breaks (Makar, 1999a). Typically, this type of failure occurs in small diameters and is caused by bending forces applied to the pipe with a failure crack propagating across the circumference of the pipe. Investigations of failures by NRC Canada's Institution for Research in Construction (IRC) have indicated that 90% of failures have corrosion pits or graphitization located at the fracture surface. In addition, IRC found that 95% of the failures showed evidence of multi-stage failure (Makar, 1999b).

Large diameter pipes generally have a higher moment of inertia and are less prone to circumferential failures (Rajani and Kleiner, In Press 2013). Although not common, there are recorded cases of circumferential cracking in large diameters (Rajani and Kleiner, In Press 2013). The relationship of circumferential to longitudinal failure modes for small (<15 in.) diameter cast iron pipe is illustrated in Table 2-3 and for all diameters in Figure 2-2. This was developed from 72,000 UK water systems data records of burst failures in the period from 1992 to 1998 collected by UK Water Industry Research (UKWIR) (Marshall, 2001). In general, circumferential failures were more prevalent in smaller diameter mains, and, by comparison, longitudinal failures were more common than circumferential failures for mains 10 in. and larger.

2.3.3   Mixed Fracture. Mixed fractures take two forms as illustrated in Table 2-2. One form is principally longitudinal cracking where a crack initiating at mid-point propagates in either or both directions depending on where soil support is least. The failure mechanism is tensile hoop stress in combination with bending.

The shattering form of failure is due to the annealing process that was used on spun cast pipes with metal molds and takes the form of multiple cracks. In this case, the failure mechanism is hoop tensile stress in combination with residual stress.

2.3.4   Bell Splitting. Bell splitting can take two forms: (1) a longitudinal crack at the bell end (Rajani and Kleiner, In Press 2013) or bell shard (Moser, 2008). The former failure mechanism is due to hoop stress and the latter is caused by hoop and flexural tensile stress. Large diameter gray cast iron pipes can fail by having a wedge section or shard of the bell shear off as shown in Table 2-2. Rajani and Kleiner suggest bending forces are more likely to be the cause of this type of failure where a wedge is split off to relieve the bending stresses and produce the type of shearing shown in the figure. Fatigue crack growth is now also suggested as a possible failure mechanism. For fatigue failure to occur there has to be a pipe defect such as a crack or manufacturing flaw, which may go undetected for years. Rajani and Kleiner (In Press 2013) performed a failure analysis on a 30 in. cast iron main that failed in Cleveland in 2008 due to a bell split. The most likely failure scenario was determined to be due to additional rotation of the pipe joint, which likely induced a small crack in the bell that grew uncontrollably under fatigue loading and eventually causing the bell to split.

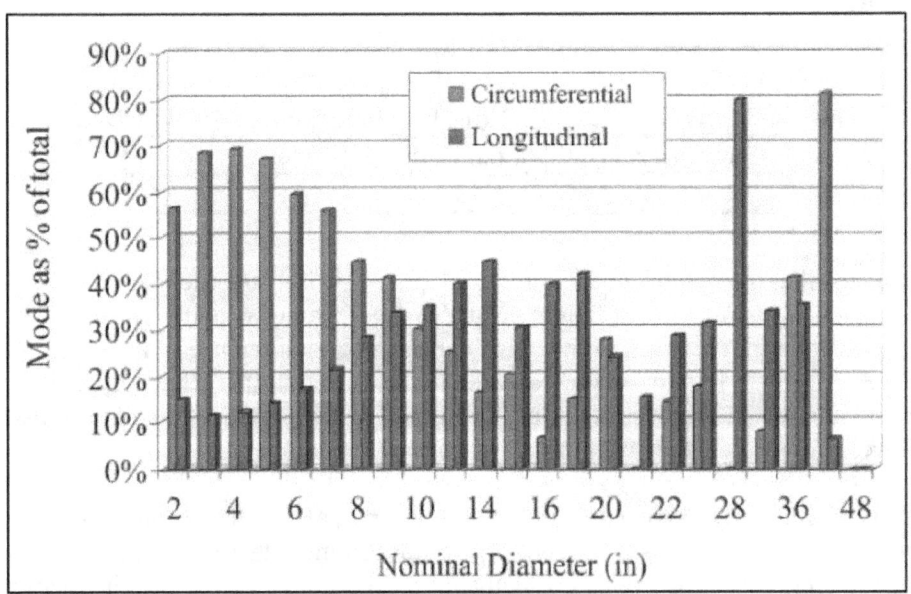

Figure 2-2. Graph of Failures Modes to Diameter (UKWIR)

2.3.5   Corrosion. Corrosion in the form of pitting and/or graphitization is a common but not exclusive factor in most pipe failures. Possible causes are localized corrosion cells, adverse soil chemistry, and bacteria. Pitting is the most common form and occurs quite randomly and leads to leaks rather than structural failures. Corrosion pitting thins and weakens the pipe wall to the point where the water pressure blows out the remaining, very thin pipe wall. This type of corrosion failure may produce a very small hole or a large one, depending on how localized the corrosion process has been and the pressure experienced by the pipe.

Wall thinning can also make the wall susceptible to failure from external loads (e.g., live loads, traffic loads, bending loads, etc.), but these loads are relatively small compared to internal pressures. Where the through wall perforation is small, the pipe does not structurally fail or in some cases even leak as the corrosion product can act as a stopper in the pipe wall hole (Marshall, 2000).

Graphitization, which is an important form of failure, is a corrosion process that removes some of the iron leaving a matrix of graphite flakes held together by iron oxide. Graphitization is often not discernible to the eye as it forms a substance with some strength albeit considerably reduced and with the appearance of normal cast iron.

2.4   Potential Contributory Factors to Failure

Pipe condition is the cumulative effect of many factors acting on the pipe (Table 2-4; Al-Barqawi and Zayed, 2006). These factors are classified into three categories: physical, environmental, and operational. The factors in the first two classes could be further divided into static and dynamic (or time-dependent). Static factors include pipe material, pipe geometry, and soil type, while dynamic factors include pipe age, climate, and seismic activity. Operational factors are inherently dynamic.

Table 2-4. Factors Contributing to Water System Deterioration

Physical Factors	Environmental Factors	Operational Factors
Pipe age and material	Pipe bedding	Internal water pressure
Pipe wall thickness	Trench backfill	Transient pressure
Pipe vintage	Soil type	Leakage
Pipe diameter	Groundwater	Water quality
Type of joints	Climate	Flow velocity
Thrust restraint	Pipe location	Backflow potential
Pipe lining and coating	Disturbances	Operation and maintenance (O&M) practices
Dissimilar metals	Stray electrical currents	
Pipe installation	Seismic activity	
Pipe manufacture		

Al-Barqawi and Zayed, 2006.

Many of the factors are not readily measurable or quantifiable, and the quantitative relationships between these factors and pipe failures are not completely understood.

2.4.1   Physical Factors

2.4.1.1   Pipe Age, Material, and Manufacture. Manufacturing defects can play a large role in the failure of pipes including non-uniform wall thickness with the earliest forms of manufacturing being the most variable. CI pipes were basically manufactured by two systems, namely pit casting and centrifugally, or spin casting. Vertical pit casting was the preferred method of manufacturing due to the longer sections of pipe that could be cast. Horizontal casting was limited by the flexural rigidity of the mold core where bending of the core could cause inconsistent wall thickness along the length of the pipe.

In the 1920s, the process of centrifugally casting gray cast iron pipe was introduced and became the primary manufacturing method of cast iron pipe by the early 1930s. This method involved pouring the molten iron into a mold that was horizontal and spinning. The speed of the spinning depended on the

9

required thickness of the pipe. In turn, cast iron pipe was gradually replaced by ductile iron pipe starting in the 1950s. Cast iron pipe was being manufactured and installed until the 1970s, albeit in limited quantities, so by definition even the most recent spun iron pipelines are 50 years old.

Probably the most common manufacturing defect is the inclusion of air trapped in the metal during the casting process. Other common manufacturing defects are unintentional inclusions of material such as ferrosilicon or iron oxide during casting, which can act as crack formers. The addition of phosphorus during the casting process to lower the melting point and viscosity can produce an iron phosphide compound which is more brittle than cast iron and weakens the pipe. Spun cast pipes have fewer casting defects than pit cast and variations in wall thickness are less. The spin casting process can produce surface flaws such as laps, laces, and pinholes. These flaws are due to the uneven cooling process that can trap air creating boundary layers that can lead to fissure cracking.

With numerous producers of pit and spun cast iron pipes using varying feed stock there are often significant differences in the micro-structure and quality. An American Water Works Association Research Foundation (AWWARF) report "Investigation of Grey Cast Iron Water Mains to Develop a Methodology for Estimating Service Life" (Rajani et al., 2000) provides detailed data on mechanical tests and metallurgical analyses properties. It is noted that the tests showed that the tensile strength and fracture toughness are significantly lower for pit than spun grey cast iron. This is attributable to the microstructure of the metal. Carbon grey flakes act as crack formers. The larger flakes in pit cast pipe make it easier for cracks to initiate.

Pit cast and spun cast iron exhibit brittle behavior. A typical stress-strain curve for both pit cast and spun cast iron together with ductile iron is shown in Figure 2-3 (from Cassa, 2008 and Sears, 1964). The brittle behavior of both pit cast and spun cast iron is apparent in the figure, with rupture failure occurring at a low value (<1%) of axial strain. The pit cast iron, which fails at a slightly lower axial strain, is slightly more brittle than the spun cast iron. Ductile iron is, as its name suggests, substantially more ductile. It has a yield point at an axial strain near the breaking strength of pit cast and spun cast iron, but then it undergoes plastic strain until it fractures at an axial strain in the low percent range, e.g., 4.5%, as in the figure. Rajani et al. (2000) has test results for failure strain for nearly 200 test samples of pit and cast iron. The variations in mechanical properties of cast iron are discussed in considerable detail in Chapter 2 of Rajani and Kleiner (In Press 2013), with spun iron having better and more consistent properties than pit cast pipes.

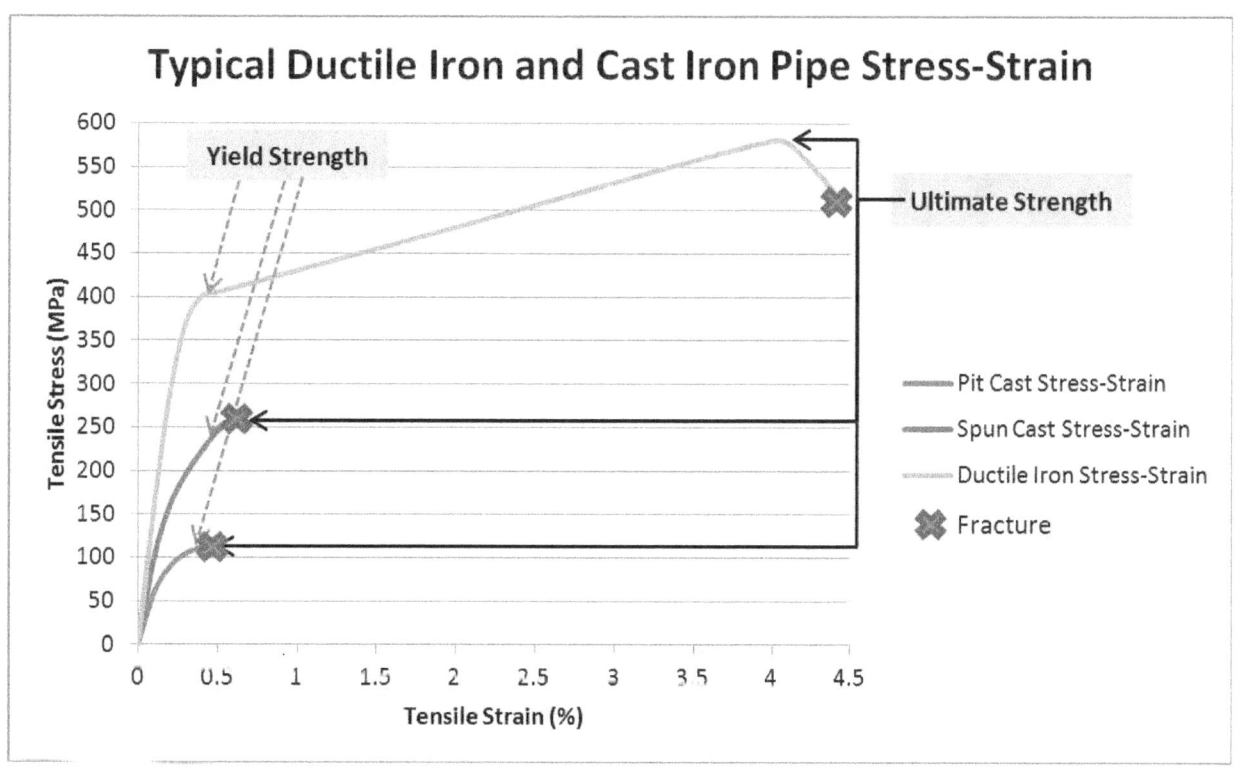

Figure 2-3. Typical Stress vs. Strain Relationship for Cast Iron and Ductile Iron

2.4.1.2 ***Pipe Wall Thickness and Vintage.*** The wall thickness of pit cast iron pipes can frequently vary around the circumference of the pipe. This can be up to 30% plus and minus variation from the average wall thickness. Spun pipes are less variable but may have some wall thickness variation. Pipe wall thickness over time has reduced for the same pressure rating, as shown in Table 2-5.

Table 2-5. Changing Wall Thickness for a 36-in. Pipe Operating at 150 psi

Year	Material	Wall Thickness (in.)
1908	Cast iron	1.58
1952	Spun iron	1.22
1957	Spun iron	0.94

In any evaluation it is important to know the pipe vintage and the original wall thickness. The vintage can assist in determining the method of production, which is important in understanding the physical and mechanical properties and potential manufacturing defects. Early cast iron foundries were numerous and set up to serve local markets. Considerable quality variations have been identified. Original pipe thickness is important in setting the baseline for determining loss of metal.

2.4.1.3 ***Pipe Diameter.*** In terms of failure rates, diameter is a significant factor and typically the larger the diameter the lower the failure rate. Based on a set of UK failure records for cast iron pipes spanning 60 years, an analysis showed the average failure rates by diameter (Table 2-6). As can be seen, the average failure rate for a 6 in. pipe is five times greater than for a 21 in. pipe.

11

Table 2-6. Failures per km/yr. by Diameter

Diameter (in.)	Failures/km/year
6	0.204
8	0.141
10	0.105
12	0.083
15	0.062
18	0.049
21	0.040

2.4.1.4   *Type of Joints*. The types of joints commonly found in cast iron pipes are:

- Bell-spigot jointed with lead
- Bell-spigot jointed with leadite (a sulfur based compound)
- Push joint sealed with a rubber gasket

A typical bell-spigot joint configuration for cast iron pipes is shown in Figure 2-4 (Rajani and Kleiner, In Press 2013). Joints in cast iron pipes were originally sealed using rope packed between the bell of one pipe and the spigot of the other. Molten lead was then poured into the joint to complete the seal. The behavior of these joints is highly dependent on the type and condition of the packing and caulking materials. In water mains, where the packing remained pliable and deformable, rotation is allowed at the joint. The packing also swelled with the absorption of water, which helps in sealing any points of leakage.

Failure at a joint can be partial or complete. Partial failure occurs at lead caulked joints when the rotation is enough to cause an opening to form in the joint where the packing and the pipe are in contact. This allows leakage to occur at the joint. Complete failure occurs when the rotation at the joint is great enough to cause the spigot to force the packing out of place and for the spigot and bell to come into contact. Leakage can also result from undetected cracks in bell and/or spigot, which may have occurred as described in 2.4.1.6.

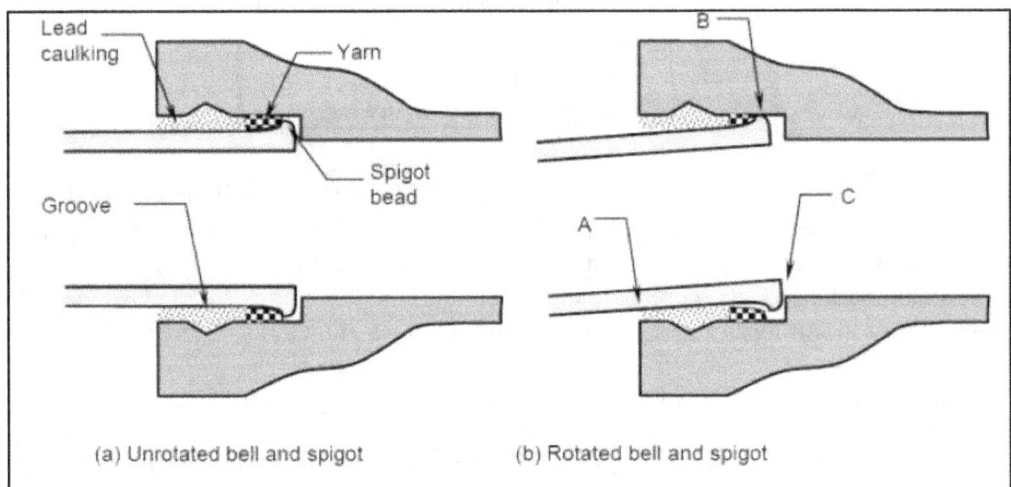

Figure 2-4. Typical Bell-Spigot Joint Configuration

This metal-to-metal contact between the two pipe sections, called metal binding, is the point where the joint is considered to have completely failed because the addition of a small rotation will cause a large increase in the stress within the joint. It has been noted that spigots that have had the asphaltic coating

12

removed can develop significantly large moments compared to joints with asphaltic coatings (Rajani and Kleiner, In Press 2013). Other noted problems include overzealous caulking particularly with pneumatic hammers, and use of oversized tools which can cause bell cracks.

Leadite, a rigid, sulfur based compound was used in the 1930s and 1940s as a substitute for lead. As a non-metallic compound, leadite has a different thermal coefficient of expansion than lead and it was a widely held belief that for very cold temperatures it was a cause of bell splitting. Rajani and Kleiner carried out a detailed test program on leadite using a temperature range of -20° to +20° C. No significant coefficient changes were noted over the range and the study demonstrated that a difference in thermal coefficients between lead and leadite is not significant enough to confirm that leadite joints are more prone to failure (Rajani and Kleiner, In Press 2013).

In the same study, Rajani and Kleiner also undertook an in-depth study of rotation limits in cast iron lead and leadite joints. They developed bi-linear and non-linear models based on limited data, much of which related to smaller diameters. They determined that a joint rotation of 0.5° is likely to allow leakage. The analysis showed that the extent to which a pipe joint can rotate without failure when subject to soil movement decreases with an increase in diameter. They also noted that the likelihood of perfect alignment when installed was unlikely, so the need is to monitor the relative changes in joint rotation rather than the absolute, as a relatively small change could put the joint in an overstressed position.

2.4.1.5   *Pipe Lining and Coating*. Cement mortar lining is used for potable water lines not only for potability, but also for inhibiting internal corrosion. Damage to any coating of the pipe wall can affect the performance of the pipe by facilitating corrosion and degradation of the pipe wall. Unlined pipe is susceptible to tuberculation and internal corrosion. Cement lining may be degraded by carrying water with a low pH or abrasion due to high water velocity and sediments.

2.4.1.6   *Pipe Installation.* A significant number of pipe failures can be attributed to the original installation. A number of reports, some going back to the early twentieth century, have drawn attention to the potential for cracks to have been created because of the way pipes were loaded, transported, unloaded, stored, and installed. There are numerous references, one dating back to 1911, which are listed and reviewed in Chapter 1 of "Fracture Failure of Large Diameter Cast iron Water Mains" (Rajani and Kleiner, In Press 2013). Small undetected cracks both in the bell and spigot can be created, which can fail over time due to fatigue from external loads and internal pressure. Detection of these cracks that have not surfaced with inspection technologies could potentially indicate a pipe in distress.

2.4.1.7   *Other Physical Factors*. Other factors include thrust restraint and dissimilar metals (see Section 2.4.2.3). Inadequate thrust restraint can increase the longitudinal stresses in the pipe, which can lead to circumferential fractures.

2.4.2   Environmental Factors

2.4.2.1   *Pipe Bedding and Backfill.* Many early installations did not take into account the need for proper bedding and backfill and practices such as supporting pipes in the trench on blocks of wood were used. The first AWWA standard issued in 1908 (AWWA 7C.1) did not address how pipes should be laid. CIPRA's pipe manual of 1927 indicated the need for continuous support and avoiding sharp or hard objects under the pipe. The manual also emphasized the need to compact trench backfill. During the 1920s, research was conducted at the University of Illinois and Iowa State on the importance of bedding and backfill. This work was fundamental to the standards that were developed over the next decades.

2.4.2.2   *Soil and Groundwater.* Depending on the way they were installed, pipes may be surrounded by natural occurring soil or by imported material. In general, soil and groundwater are not very aggressive and correlations between soil type and deterioration can be unreliable and tenuous. However, there are exceptions where aggressive and contaminated soils are involved, and external corrosion can be created. There is a great deal of literature from researchers and vendors proposing relationships between soil properties and corrosion: Ductile Iron Pipe Research Association (DIPRA, 2005); Ferguson and Downey (2009); Booth et al. (1967); and Jarvis and Hedges (1994).

2.4.2.3   *Galvanic Corrosion.* Galvanic corrosion can occur when dissimilar metals are electrically connected. This is more likely to occur in distribution mains where connections are made onto the pipe. A galvanic cell can also occur when a CI pipe is installed in a non-uniform soil. An example would be where lumps of clay are in contact with the pipe in a sand backfill. Another factor contributing to corrosion is pipe location (e.g., the migration of road salt into the soil can increase the corrosion rate).

2.4.2.4   *Stray Electric Currents.* Stray electric currents are generated by an adjacent direct current (DC) source. A ferrous pipe can offer a better earthen route for conveying stray currents from electrified transport systems, electrical installations such as pylons, and cathodic protection systems. Most cast iron installations are not electrically continuous as lead and leadite are poor electrical conductors. Pipes jointed with rubber gaskets are also not considered to be electrically continuous. A number of authorities have retrofitted lines with impressed current or sacrificial anodes to achieve cathodic protection of pipes.

2.4.2.5   *Microbiologically Influenced Corrosion.* The two forms of microbiologically influenced corrosion (MIC) are anaerobic and aerobic. Sulfate reducing bacteria (SRB) are typical examples of anaerobic bacteria. Corrosion can occur even in the absence of dissolved oxygen (Ferguson and Nicholas, 1984).

2.4.2.6   *Soil Movement and Disturbances.* When pipes are laid in trenches, they are subject to external pressures that act on the pipe. These external loads are not symmetrical; therefore, the differential loading can cause bending in the pipe wall. Possible loads on buried pipes include vertical soil pressure, superimposed live loads due to vehicles, frost loading, self weight of pipe and its contents, crushing or bending by heaving, swelling, or contraction of soils, and even seismic activity causing increased stresses on the pipe. Thus, the excessive loads can result in failure due to either crushing or compression in the pipe.

Failure of cast iron pipe with bell joints was reviewed earlier in this section. Soil movement can cause joint rotation, leading to leakage and ultimately failure. The soil movement can be due to external causes or by joint leakage eroding bedding support. It is quite common to find that leakage has gone undetected for long periods. Table 2-7 provides information on the potential for loss of support, soil movements, and soil expansion.

Table 2-7. Soil Types and Impact on Structural Defects

Soil Type	Potential for Loss of Support	Potential for Soil Movements	Potential for Soil Expansion
Rock	Low	Low	Low
Gravel above water table	Low	Low	Low
Gravel below water table	Moderate	Low	Low
Sand/silts above water table	Moderate	Moderate	Low
Sand/silts below water table	High	High	Low
Clay	Low	Low	High
Organic	High	High	High

2.4.2.7 *Climate.* Changing soil temperatures are of concern in northern regions where freeze/thaw cycles can cause soil movement. The impact appears to be greatest on small diameters. Precipitation is another factor, as it impacts soil moisture, which in turn impacts frost penetration and expansion and shrinkage of expansive clay soils.

2.4.3 **Operational Factors**. Poor or incorrect operational and maintenance practices can be a factor in the failure process.

2.4.3.1 *Hydraulic Factors.* Hydraulic operational factors that can influence failures include internal water pressure, transient pressures, and water hammer. Large and rapid changes in flow velocity (e.g., fast valve closures or power outages in pumping stations) can create transients greater than the design limits. High pressures create high stress in the pipe and a higher likelihood of failure. Frequent pressure changes can be a factor in fatigue failure. A combination of loss of metal or flaws and transient pressures can be a significant cause of failure.

2.4.3.2 *Leakage.* Leakage is common and often goes undiscovered. As discussed in Section 2.4.1.4, as little as 0.5° of rotation will lead to leakage from a bell and spigot lead or leadite joint. This leakage can erode support and allow further rotation and failure.

2.4.3.3 *Other Operational Factors*. Other factors include: water quality (e.g., aggressive water promoting corrosion); flow velocity (e.g., unlined dead-ended water mains having a higher rate of corrosion); backflow potential from cross connections with non-potable water systems; and poor O&M practices.

2.5   Condition Assessment - Distress and Inferential Indicators

Many of the factors contributing to failure are not readily measurable or quantifiable, nor are the quantitative relationships between these factors and pipe failures always well understood. In undertaking a condition assessment program and undertaking a condition investigation, the use of distress and inferential indicators can greatly aid the process.

Distress indicators are defined as the observable/measurable physical manifestations of the aging and deterioration process. Distress indicators are a result of some or all of the factors listed above. Each distress indicator provides partial evidence for the condition of specific pipe components.

Inferential indicators (U.S. EPA, 2012a) point to the potential existence of a pipe deterioration mechanism. They do not provide direct evidence, but rather indicate the possibility without knowing if this potential has actually been realized. Environmental indicators, such as soil type, groundwater fluctuations, etc., are inferential in nature. These indicators are cost effective in pre-screening pipes to select those that should receive more expensive direct inspection.

2.5.1 **Distress Indicators for Cast Iron Pipe**. Distress indicators as set out in Table 2-8 can be discerned by direct observation in some case, while many require more sophisticated methods of investigation (Kleiner et al., 2005). It is important to understand that the information obtained will come from a variety of sources and forms and needs to be aggregated and interpreted by some kind of pipe condition rating system to quantify the condition.

Table 2-8. Distress Indicators that Influence Pipe Condition for Cast Iron Pipes

Category	Distress Indicator	Comments
External coating (poly wrap/tar/etc)	Crack/tear/holiday	State of external coating will dictate how external corrosion is likely to encourage damage to the pipe.
External pipe barrel/bell	Remaining wall thickness	Remaining pipe wall thickness is usually obtained from nondestructive evaluation (NDE) tests or from spot exhumations and sand blasting samples. Casting defects (voids or inclusions) can be of significant size in CI pipes.
	Graphitization (pit) aerial extent	Areal extent as percentage of pipe diameter times unit length indicates the size of affected area. Severe graphitization may not always mean the pipe should have failed. In practice, graphitized area can still provide some resistance – it acts as a form of sticky plaster. In CI graphitization is typically in the form of graphite flakes.
	Crack (pit) type	Circumferential cracks indicate some type of longitudinal movement, loss of bedding support, or increase in vertical load (e.g., frost) has taken place. Longitudinal cracks occur due to low hoop resistance and can be caused by internal water pressure, external vertical loads, or compressive forces acting along the pipe.
	Crack (pit) width	Crack width is another indicator of corrosion. A wide crack together with a deep pit will be more detrimental to the pipe than a narrow but shallow crack.
Inner lining/ surface	Cement lining spalling	Inner lining deterioration is often due to incompatible water chemistry or abrasion due to the presence of high water velocities and sediments.
	Remaining wall thickness	Occasionally closed circuit television (CCTV) scans can indicate internal corrosion pits. NDE testing is required to give qualitative information on remaining wall thickness.
	Tuberculation	Heavy tuberculation can hide defects in the wall including pitting. It also needs to be removed for internal inspection by NDE tools. It can significantly reduce water delivery and produce red water condition.
Joint	Change in alignment	Changes in joint alignment (rotation) indicate pipe susceptible to ground movement. As little as 0.5° or rotation can lead to leakage and eventually joint failure (Rajani and Kleiner, In Press, 2013).
	Joint displacement	Joints can displace without undergoing joint misalignment and hence is also an indicator of other forces at play.

2.5.2   Inferential Indicators for Cast Iron Pipes. Inferential indicators are a pointer to a potential cause or causes of failure without providing any direct information on pipe deterioration. Some indicators can be found in data records, while others are environmental and will be specific to the location. They are valuable as a pre-screening tool and relatively easy and cheap to use. The inferential indicators described in Table 2-9 are adapted from Kleiner et al. (2005).

Table 2-9. Inferential Indicators for Cast Iron Pipes

Category (Level 1)	Agent (Level 2)	Comments
Pipe vintage	Material type	Pipes of specific vintages have experienced a higher breakage rate.
Water quality	Water pH	Water with low pH can leach the internal cement lining or pipe wall itself if lining is absent.
Water loss	Leaks	May be due to through wall pitting corrosion or joint displacement
Water pressure	Operating pressure (OP)	High pressure subjects pipe to high stress and hence higher propensity to failure.
	Pressure change amplitude (% OP)	Large pressure changes (% of OP) can induce higher stresses than expected by design.
	Pressure change frequency	Either slow or fast fatigue mechanism can induce early failure.
Location	Pipe embedment	Pipes exposed to wet/dry conditions have higher failure rate than pipes totally below water table or pipes totally exposed to atmosphere.
	Surface loads - traffic type	Heavy surface loads will subject the pipe to high stresses and hence faster deterioration in the long term.
	Wet/dry cycle(s)	Changing environment can promote corrosion.
	Water table level	Water table position will indicate if wet/dry cycle is likely to occur.
Soil	Soil type / backfill	Non-draining backfill leads to moisture retention and hence promotes corrosion.
	Soil resistivity (ohm-cm)	Low resistivity soil leads to higher corrosion rates
	Soil pH	Low pH (< 4) means soil is acidic and likely to promote corrosion; high alkaline conditions (pH > 8) can also lead to high corrosion.
	Redox potential	High availability of oxygen promotes MIC in the presence of sulfates and sulfides.
	Soil chloride	Low chloride levels in high pH (> 11.5) environments can lead to serious corrosion.
	Soil sulfate	Accounts for MIC and possible food source for SRB in anaerobic conditions under loose coatings.
	Soil sulfide	Sulfate reducing bacteria give off sulfides, which are excellent electrolytes.
	Frost susceptibility (load)	CI pipes are not designed for frost loads. If conditions exist to develop significant frost loads, then pipe will be subjected to additional stresses (annual) and lead to pipe failure if already significantly corroded. These conditions are: high water table; thermal gradient; right soil type to develop suction (i.e., silt or clayey silt).
Corrosion	Cathodic protection	Impressed current is likely to reduce corrosion.
	Stray current	Stray current is known to accelerate corrosion unless adequate measures have been taken.

SECTION 3.0: STRUCTURAL INSPECTION COMPONENTS AND SYSTEMS

3.1 Overview

To successfully monitor structural condition, a combination of screening, monitoring, and condition assessment techniques need to be used. This section briefly presents available inspection and monitoring technologies with the potential for detecting and monitoring pre-failure damage indicators in large diameter cast iron mains (greater than 16 in. in diameter). The broad categories of available inspection technologies are presented, along with a summary of the types of defects and pre-failure indicators that can be detected. For each category of inspection tool or component, the applicability and/or known limitations for the inspection of large diameter cast iron water mains are also noted.

The purpose of the protocol development in this report is to identify and evaluate emerging inspection and monitoring technologies, including those developed for other applications, and to assess their applicability and suitability for large diameter cast iron water mains. For that reason, this section also documents the organizations involved in structural inspection technology and component research. Although this report only provides a snapshot of the research currently being conducted for structural inspection components and systems, several of the key stakeholders involved in inspection technology research remain fairly constant over time including certain Federal agencies and non-profit research organizations. In addition, private companies and technology vendors are an important source of fundamental research and technology development of inspection tools and platforms. These key stakeholders and their overall research interests are summarized here in order to aid in the identification and screening of emerging structural inspection technology research for cast iron water mains discussed later in this report.

3.2 Available Inspection and Monitoring Technologies

This section provides a broad overview of structural inspection technologies with some observations on their suitability and known limitations for cast iron water main inspections. More detailed reviews of these technologies can be found in the EPA reports titled, *Condition Assessment of Ferrous Water Transmission and Distribution Systems* (U.S. EPA, 2009) and *Condition Assessment Technologies for Water Transmission and Distribution Systems* (U.S. EPA, 2012a). In addition, numerous sources provide detailed information on inspection and monitoring technologies including: American Water Works Research Foundation (AWWARF) reports titled, *Techniques for Monitoring Structural Behavior of Pipeline Systems* (Reed et al., 2004) and *Workshop on Condition Assessment Inspection* (Lillie et al., 2004); the Water Environment Research Foundation (WERF) reports titled, *Inspection Guidelines for Ferrous Force Mains* (Jason Consultants, 2007) and *Condition Assessment Strategies and Protocols for Water and Wastewater Utility Assets* (WERF, 2007); *Control and Mitigation of Drinking Water Losses in Distribution Systems* (U.S. EPA, 2010); Rajani and Kleiner (2004); and the UKWIR report titled, *A Survey of Practices for the Detection and Location of Leaks* (UKWIR, 2011).

3.2.1 External Inspection Technology Description. External condition assessment tools provide detailed condition information for selected locations along the pipeline and then rely on statistical methods to predict the condition of the entire pipeline segment. Often the detailed external assessments are supplemented with soil corrosivity and coating condition data to improve confidence in the statistical predictions. Although these technologies are capable of inspecting the entire pipeline length, excavation is required but it is rarely practical.

External inspection of pipelines has been widely used because it allows the pipeline to remain in service while the localized condition of the pipe is being assessed. Often small areas around the pipe must be cleared because the sensors on the device need to have contact with the outside pipe surface. Table 3-1 summarizes the overall applicability of structural inspection tools for external inspection of cast iron water mains and the type of defects that can be detected.

Table 3-1. Tools and Technologies for Inspecting Structural Integrity Externally

Application	Pit Depth Measurement	Ultrasonics	MFL[b]	(BEM)[c]
Diameters	Any	Any	≥ 6 in.	≥ 2 in.
Typical Length Scanned	3 to 6 ft	3 to 6 ft	3 to 12 ft	3 to 12 ft
Line in Operation	Yes	Yes	Yes	Yes
Scan through Coatings	No	No	Yes	Yes
Scan through Pipe Wall	No	Yes	Yes	Yes
Loss of Metal	Yes	Yes	Yes	Yes
Pit Depth	Yes	No	No	Yes
Graphitization	No	No	Yes	Yes
Cracks	No	No	Yes	Yes
Mobilization Costs	Low	Low	Med	Low
Scanning/Processing Cost	Low	Med	Med	Low/Med[a]
Suitable for Water Mains	Yes	Yes	Yes	Yes

[a]Real time provides immediate condition, while full data processing is an additional cost.
[b]MFL = Magnetic Flux Leakage. [c]BEM=Broadband Electromagnetic
Adapted from U.S. EPA, 2009.

3.2.2   Internal Inspection Technology Description. Inline inspection technologies have been used for years in the oil and gas industry to inspect pipelines for structural integrity issues such as corrosion and mechanical damage. Inline inspection technologies for water mains can range from relatively simple closed circuit television (CCTV) visual tools that assess the inner diameter of the pipe to complex tools that assess the pipe wall thickness including remote field technology (RFT) tools. The more complex technologies have only recently been used by utilities for inspection of large water mains after a few main breaks that resulted in extensive service disruptions, significant property damage, and costly repairs. Inline inspection systems provide valuable pipeline condition information especially for water mains that cannot be taken out of service.

Issues that must be overcome for wide-spread use of inline inspection technologies for water mains includes the lack of launching and receiving facilities on existing water mains, the variety of materials used to construct water pipelines, and the expense of conducting such inspections. Table 3-2 summarizes the overall applicability of structural inspection tools for internal inspection of cast iron water mains and the types of defects that can be detected.

Table 3-2. Tools and Technologies for Inspecting Structural Integrity Internally

Application	Man-Entry	CCTV	RFT	BEM
Diameters	≥ 24 in.	≥ 4 in.	up to 28 in	> 6 in.
Typical Length Scanned	Any	500 ft	10,000 ft	3,000 ft
Line in Operation	No	No	Possibly	No
Scan through Linings	No	No	Possibly	Yes
Loss of Metal	No	No	Yes	Yes
Pit Depth	Yes	No	No	No
Graphitization	Yes	No	Yes	Yes
Cracks	Yes	Possibly	Yes	Yes
Mobilization Costs	Med	Med	Med	Med
Scanning /Processing Cost	Low	Low	Med	Med/High
Suitable for Water Mains	Yes	Yes	Yes	Yes

Adapted from U.S. EPA, 2009.

19

Internal MFL technology was not included in Table 3-2, but the use of this technology for large diameter water mains is currently being explored (Hannaford et al., 2010).

3.2.3   **Leak Detection Technology Description**. The main objectives of leak detection are the reduction (or elimination) of water losses through leaks, as well as reducing the possibility of small leaks developing into pipe failures. While addressing these two main objectives, information about leakage rates provides an important indication about the condition of the pipe (U.S. EPA, 2012a,b). Table 3-3 summarizes leak detection technologies suitable for cast iron water mains.

Table 3-3. Tools and Technologies for Leak Inspection

Application	Visual Inspection[a]	Leak Correlators	Listening Sticks	Acoustic Leak Detection
External/Internal	External	External	External	Internal
Diameters	Any	Most	Most	$\geq$ 12 in.
Typical Length Scanned	Any	300 ft	3 ft	Miles
Line in Operation	Yes	Yes	Yes	Yes
Joint Leaks	No	Yes	Yes	Yes
Wall Perforation Leaks	No	Yes	Yes	Yes
Accuracy Locating Small Leaks	Poor	Good	Fair	Excellent
Insertion into Line	N/A	N/A	N/A	via valve or tap
Mobilization Costs	Low	Low	Low	Low/Med
Scanning/Processing Cost	Low	Low	N/A	Low/Med
Suitable for Water Mains	Yes	Yes	Possible	Yes

[a] Can detect leakage or ground movement, but not leak type or location.

3.2.4   **Summary**. Table 3-4, which is adapted from the 2009 and 2012 EPA reports, focuses the structural inspection technology discussion on the advantages and limitations of the technologies suitable to large cast iron water mains (>16 in. diameter).

3.3   Structural Inspection Technology Research Applicable to Cast Iron Water Mains

Key U.S. stakeholders involved in structural inspection technology research are briefly mentioned here to serve as a preliminary guide to potential sources of emerging technologies that could be evaluated for their suitability for large diameter CI water mains. Research from multiple Federal organizations has been reviewed including the: EPA; U.S. Department of Transportation (DOT); U.S. Department of Energy (DOE); U.S. Department of Defense (DOD); U.S. Department of Commerce (DOC); U.S. Department of Homeland Security (DHS); U.S. Department of the Interior (DOI); National Science Foundation (NSF); and National Aeronautics and Space Administration (NASA). Also, research from non-profit research organizations such as WaterRF, WERF, and the Gas Technology Institute (GTI) as well as private industrial research is briefly discussed. International stakeholders, particularly from Canada, the UK, Australia, and Germany are also important sources of innovative inspection technologies and procedures.

Appendix A briefly describes the role of each agency in structural inspection technology research. Table 3-5 summarizes the organizations that fund research that is potentially relevant to structural inspection for large diameter cast iron water mains. For each organization, example research activities and projects are listed, but this is not a comprehensive list. The EPA report *White Paper on Improvement of Structural Integrity Monitoring for Drinking Water Mains* (Royer, 2005) lists other projects undertaken by these organizations for non-drinking water systems, which may have potential for application to water conveyance systems.

Table 3-4. Available Inspection and Monitoring Technologies Applicable to Cast Iron Mains

Technology	Diameters	Advantages	Limitations
External Technologies			
Pit depth measurement	All	• Direct measurement of pit depth, no need for interpretation. • Provides good indication of sample condition. • Exposed pipe does not need to be taken out of service.	• Requires statistical analysis to infer general condition of CI. • Existing coating must be removed and pipe exposed. • Thickness must be known for corrosion estimate and testing varies.
Ultrasonics	All	• Sensitive to both surface and subsurface discontinuities. • Provides instant results of metal loss. • Probes of different sizes and frequencies are available. • Supply shutdown is not necessary.	• CI and other coarse grained materials are difficult to inspect due to low sound transmission and high signal noise. • Surface to be inspected must be accessible and clean. • Coupling medium required for some products.
Magnetic flux leakage	≥ 6 in.	• Detects cracks, graphitization, and measures wall thickness. • Does not require a service interruption.	• Accuracy is higher if sensors maintain direct contact with the CI pipe wall. • Not widely used due to cost.
Broadband electro-magnetic	≥ 2 in.	• Contact with the pipe wall not required. • Scans through coatings and linings. • Detects cracks, graphitization, and metal loss.	• Resolution depends on size of the sensor. • Unable to define/quantify pin-hole failures or isolated pits. • Scanning process is not continuous.
Acoustic wall thickness	> 2 in.	• Finds average wall thickness between excavation points. • No contact with water.	• Resolution depends on spacing between excavation points. • One bad pipe length in a pipeline that is generally good may not be detected
Internal Technologies			
Man entry inspection	≥ 24 in.	• No special equipment required. • Assessment can provide an indication of the cause of the deterioration and the likelihood of being more widespread.	• Only for man-entry CI pipes. • Not effective finding defects not on the inner pipe surface. • Mains need to be taken out of service.
Broadband electro-magnetic	≥ 6 in.	• Contact with the pipe wall not required. • Scans through coatings and linings. • Detects cracks, graphitization, and metal loss.	• Resolution depends on size of the sensor. • Unable to define/quantify pin-hole failures or isolated pits. • Scanning process is not continuous.
Closed circuit TV	≥ 4 in.	• Digital recording is convenient for data storage, as well as future developments in automatic data interpretation.	• Data for inner pipe wall only and results need interpretation. • Not for in-service mains and does not find structural defects.
Remote field eddy current	≤ 28 in.	• Inspection of in-service pipes is possible. • Detects cracks, graphitization, and metal loss.	• Data interpretation needs experience. • Some tools require pipe cleaning and/or dewatering.
Internal acoustic wall thickness	> 2 in.	• Finds average wall thickness at discrete intervals as small as 1 ft and as large as tens of feet. • Has the potential to determine relative wall thickness for specific pipe lengths.	• Individual anomalies not detected. • Used internally which requires excavation and access.

Table 3-4. Available Inspection and Monitoring Technologies Applicable to Cast Iron Mains (Cont.)

		Leak Detection Technologies	
Visual inspection	All	• Reveals leakage and ground movement from the surface. • Allows for assessment of the backfill.	• Cannot detect leak type or location from the surface. • May not detect non-surfacing leaks and is costly to expose pipe.
Leak correlators	Most	• Used externally on operational lines. • Locates leaks in joints and pipe walls.	• Scan length is limited to around 300 ft. • Not as effective on transmission mains.
Listening sticks	Most	• Used externally on operational lines. • Locates leaks in joints and walls.	• Scan length is very short (~3 ft). • Difficulty locating small leaks.
Acoustic leak detection	> 12 in.	• Used internally on operational lines. • Long surveys with a single insertion. • Detects small noise disturbances.	• Requires tapping for access points. • May not see large leaks.

Table 3-5. Organizations Funding Structural Inspection Research Potentially Relevant to Water

Orgs	Example Research Activities
EPA	• Extramural research (e.g., cooperative agreements and contracts; condition assessment of ferrous water transmission and distribution systems [U.S. EPA, 2009]) • Periodic SBIR solicitations (e.g., in situ imaging of water pipelines using ultrasonics [Mu, 2011]) and STAR grants from the ORD NCER or Regional programs • NRMRL WSWRD water supply research • ETV, CEIT, and ITSC research programs • Leak detection research, pipeline test facilities, and in-house research • SDWA research (i.e., stakeholder input/assessment) of adverse water quality and health effects from distribution systems in Total Coliform Rule
DOT	• OPS; large extramural research program for natural gas and hazardous liquid pipelines, focus on 3-5 year horizon; on-line R&D project database; SBIR component; demonstration program
DOE	• Fifteen national labs, many with NDE R&D • Natural gas pipeline research - NETL; sensors - Argonne, Sandia, Oak Ridge National Labs; R&D for nuclear power (e.g., boiler tubing) and waste (e.g., pipe transport of low level waste); Intelli-Pipe to enhance data transfer from drill bit to surface
DOD	• USACE/CERL infrastructure research; RDT&E program; SERDP/ESTCP: leak detection research • Industrial Ecology Center- depot environmental management and compliance-related R&D • ARL sensors; NTIAC; Naval Facilities Command: study spill and leak prevention for pipes
DOC	• NIST TIP: projects to advance composite pipes for energy exploration and recovery; and supporting innovation through high-risk, high-reward research in areas of critical need. • TIP Project: Defect recognition using ultra wide band pulsed radar profilometry (NIST, 2011)
DHS	• Science and technology research into pipeline security monitoring
DOI	• BOEMRE/TA&R program supports operational safety and pollution prevention research
NSF	• Sponsors a broad range of basic research in relevant areas (e.g., innovative sensors, materials, NDE, information technology, data analysis); SBIR; grant programs • National Workshop on Future Sensing Systems (Glaser and Pescovitz, 2002) • No. 9901221: Non-contact sensors for pipe inspection by lamb waves (Kundu, 2005)
NASA	• LaRC research of NDE technologies and interest in new materials
WaterRF	• 2727: Effects of corrosion pitting on circumferential failures in grey cast iron pipes (Makar, 2005) • 2689: Potential techniques for the assessment of joints in water distribution pipes (Reed et al., 2006) • 4035: Fracture failure of large diameter cast iron water mains (Rajani and Kleiner, 2011) • 4234: Practical tool for deciding rehabilitation vs. replacement of cast iron pipes (WaterRF, 2011a) • 4360: Acoustic signal processing for pipe condition assessment (WaterRF, 2011b)
WERF	• 01-CTS-7: Examination of innovative inspection methods (WERF, 2004) • 04-CTS-6UR: Inspection Guidelines for Ferrous Force Mains (Jason Consultants, 2007) • 03-CTS-20CO: Condition assessment strategies for utility assets (WERF, 2007)
GTI	• Project No. 4.8.D: Broadband electromagnetic technology - sensor to measure wall thickness.

4.0: PROTOCOL DEVELOPMENT

Effective and economical structural inspection can be an important component of asset management for aging and deteriorating water conveyance infrastructure. Structural inspection provides data that can be used to support estimates of current and future structural condition of water mains. These estimates can be used to help optimize decisions about inspection, repair, rehabilitation, and replacement of water mains. The value of optimal renewal decision making arises from (1) safely utilizing installed infrastructure to its full life, (2) reduction of main break failures and their adverse health, safety, environmental, and economic effects, and (3) prompt recognition and correction of significant leakage or deterioration.

Scientific and engineering research is being conducted to develop and evaluate better, faster, and less costly inspection technologies for water mains and other applications. To accelerate commercial implementation, portions of the development and evaluation work are funded by government and industry associations. Since resources are limited, it is very desirable to focus available resources on the most useful and promising innovative condition assessment technologies. A thorough, systematic protocol for reviewing innovative condition assessment technology options would be a useful tool for making and justifying structural inspection technology research decisions. It is expected that any technology evaluation protocol will be strongly influenced by the type of pipe and its associated failure behavior. Large diameter cast iron pipe is an excellent type of pipe upon which to focus the initial condition assessment technology evaluation protocol. Large diameter cast iron pipes have been commonly installed, the consequences of failure can be high, and they are among the older pipes in the inventory. Also, their failure behavior has been studied in some detail (e.g., by Cleveland Water Department, WaterRF, NRC Canada), and the results of these studies can potentially help identify the monitoring parameters and levels that are required for successful structural inspection and remaining life estimation.

The objective of the three screening protocols described in this section is to assist research funding organizations, such as the EPA, in strategically evaluating the feasibility of emerging structural inspection technologies for large diameter cast iron mains. The first screening protocol collects the data needed to enable a user to determine if an inspection technology can be practically implemented on a large diameter cast iron water main. The second screening protocol collects the data that enables a user to determine the degradation condition or conditions that an inspection technology can detect and determines if the technology locates the key distress indicators for large diameter cast iron water mains as identified in Section 2 of the report. The third screening protocol compares a candidate technology to existing technologies and determines the potential for further development.

A user is asked to answer the protocol questions in the series of flowcharts that follow based on the current configuration of the technology. A second pass through the protocols can be performed based on the development of a technically feasible modified configuration to the technology.

4.1   Basic Screening Protocol

Protocol 1 begins (Figure 4-1) by determining suitability for large diameter cast iron pipe and general information about the intended capabilities of the technology. A reasonable large diameter minimum was determined to be 16 in. based on a survey of experienced water utilities and researchers. Conditions that may limit the suitability of a technology for inspecting large diameter cast iron mains include: tool length (e.g., some technologies require a two pipe diameter separation between source and receiver, which translates to a 4 ft separation for 24 in. pipe); large number of sensors (e.g., two sensors per inch of circumference is 25 for a 4 in. pipe and 100 for a 16 in. pipe) for which the data must be processed and stored); and launch and receive methods (e.g., cost of valves and fittings increases with diameter).

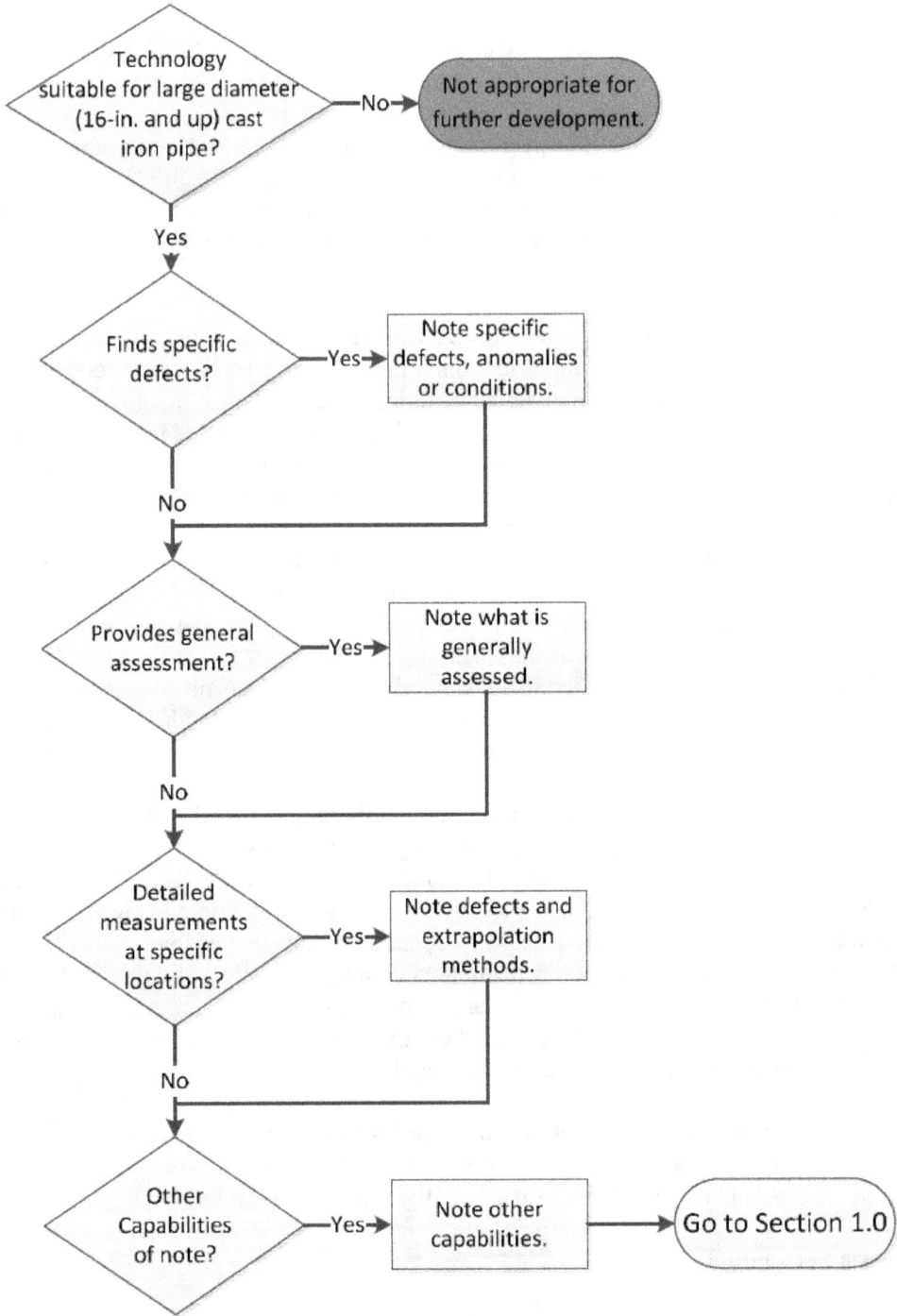

Figure 4-1. First Step of Basic Screening Protocol: Determine Technology Capabilities

Now that the intended capabilities of the technology have been noted, Section 1.0 can be used to determine the primary category of the technology.

Section 1.0: Primary Categories

The primary categories of technologies are broken down into five primary categories, determined by the flowchart in Figure 4-2. Once the correct category is selected, the user can follow the hyperlink below each box to the appropriate section.

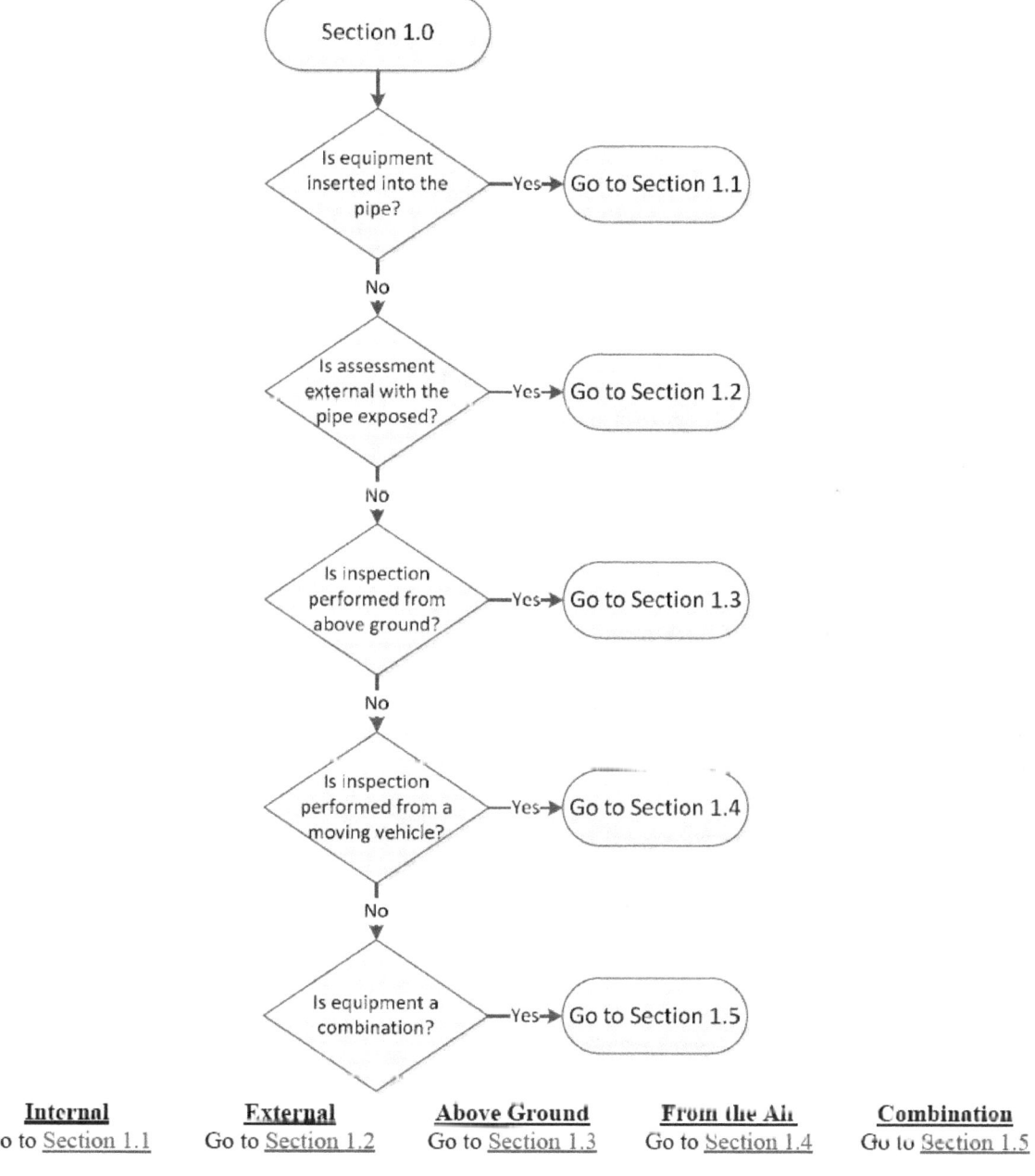

Internal	**External**	**Above Ground**	**From the Air**	**Combination**
Go to Section 1.1	Go to Section 1.2	Go to Section 1.3	Go to Section 1.4	Go to Section 1.5

Figure 4-2. Primary Technology Categories

Section 1.1: Internal Inspection

Section 1.1.1: Internal Condition. The internal inspection begins by determining the internal conditions required for the technology to be used, determined by the flowchart in Figure 4-3.

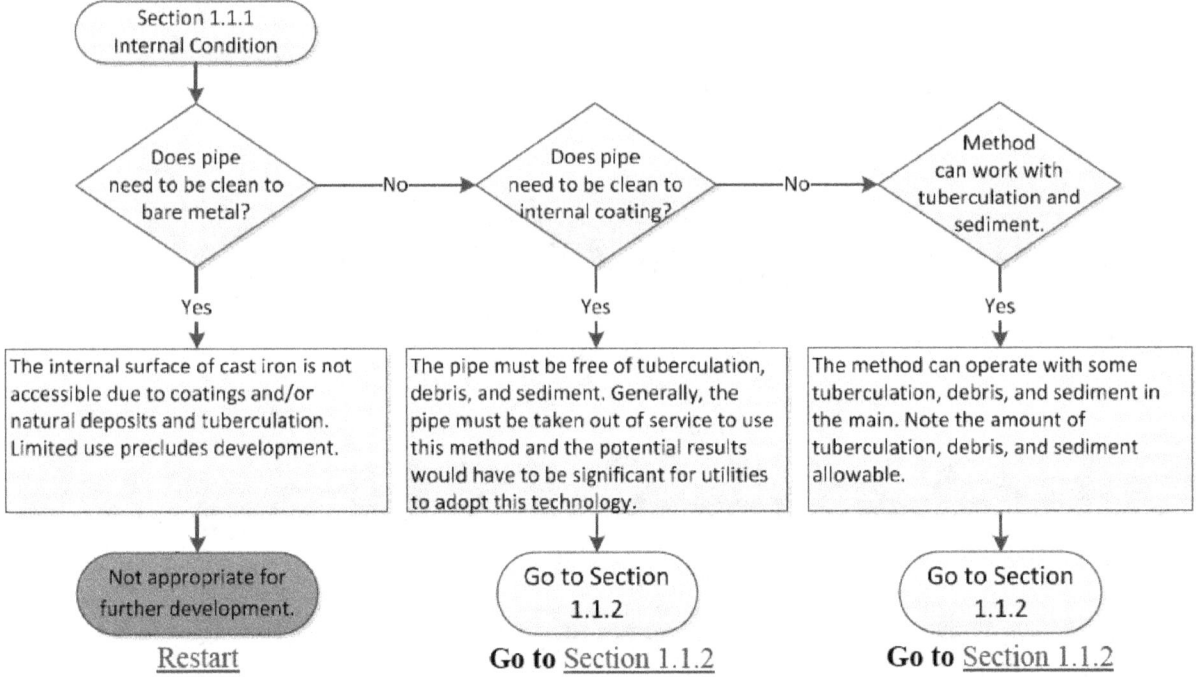

Figure 4-3. Required Internal Condition

Section 1.1.2: Implementation Questions. Implementation issues (Figure 4-4) that govern the use of structural inspection technologies are evaluated next. If a technology provides useful and detailed data, the water utility will accept some significant implementation inconvenience to use the technology. If a structural inspection technology provides more general data, the utility may still use the technology, if implementation is easy. This section collects the data needed to make an implementation decision.

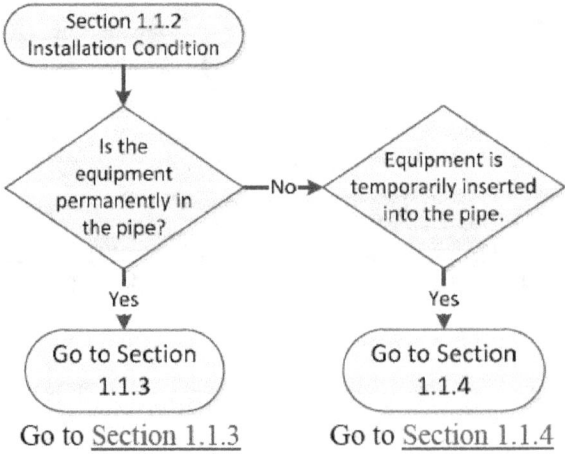

Figure 4-4. Installation Conditions

Section 1.1.3: Permanent Installation. If the method is permanently installed in the water main, the sensors used, connection method, power source, and data handling should be determined and noted, determined by the flowchart in Figure 4-5, and then Protocol 2 can be used.

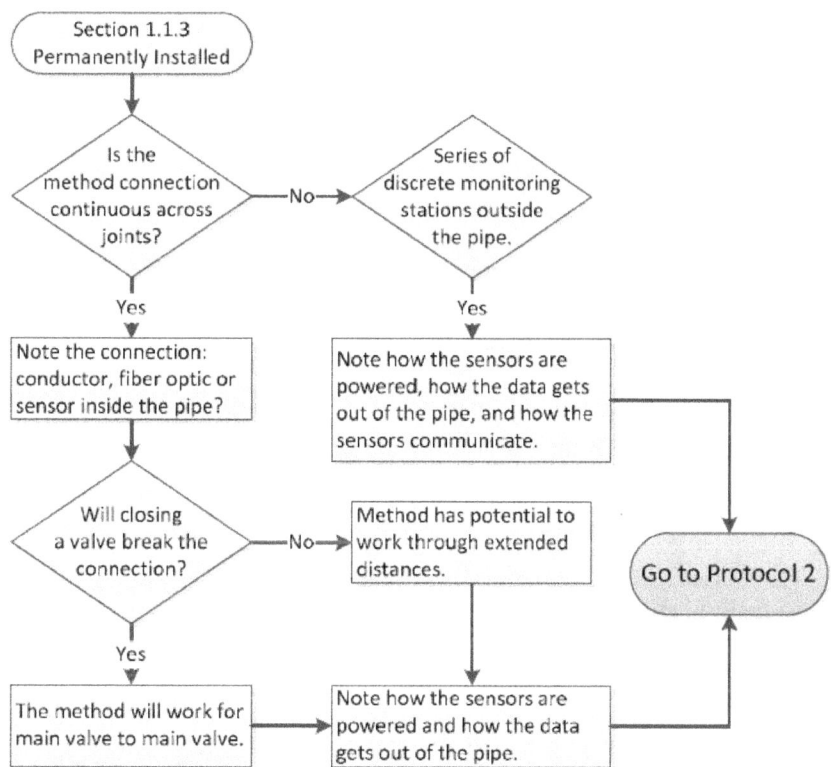

Figure 4-5. Permanently Installed

Section 1.1.4: Temporary Internal Installation. If the device is temporarily installed in the water main, the following questions should be answered and noted before going to Protocol 2.

Section 1.1.4.1: How does the device maneuver through the pipe?
    A. Free swimming, propelled by water flow
    B. Tethered, pushed or propelled by water flow
    C. Powered by robotic crawler
    D. Pulled through the main
    E. Other, note process

Section 1.1.4.2: What is the launch diameter of the method tool?
    A. Tool diameter $\leq$ 6 in.
    B. 6 in. < Tool diameter $\leq$ ½ the pipe diameter
    C. Tool diameter > ½ the pipe diameter
    D. Tool diameter is nominally the pipe diameter

Section 1.1.4.3: What is the launch angle of the method tool?
    A. Perpendicular to the pipe
    B. With a 'Y' fitting to the pipe
    C. Parallel to the pipe
    D. Other, note angle or configuration

Section 1.1.4.4: What is the receive angle of the method tool?
    A. Perpendicular to the pipe or returns to launch point for retrieval
    B. With a 'Y' fitting to the pipe
    C. Parallel to the pipe
    D. Other, note angle or configuration

Section 1.1.4.5: Can the fitting be installed while the pipe is pressurized?
    A. Yes
    B. (Intentionally blank)
    C. No

Section 1.1.4.6: How much flow is needed/allowed in the main?
    A. Line is operational, no flow restriction or flow is allowed within a specific range
    B. Line is full, but not operational
    C. Line is partially full
    D. Pipe must be taken out of operation and dewatered

Section 1.1.4.7: What is the smallest diameter of tees, branches, etc. that the flow must be stopped or barred to prevent the tool from entering the connection for the tool to work?
    A. Diameter > ¾ main diameter
    B. ½ main diameter ≤ diameter < ¾ main diameter
    C. ¼ main diameter ≤ diameter < ½ main diameter
    D. Diameter < ¼ main diameter
    E. Other

Section 1.1.4.8: Can the method pass protrusions in the pipeline?
    A. Large diameter (nominally ¼ of main diameter or greater) ½ diameter into the main
    B. Large diameter (nominally ¼ of main diameter or greater) 1 in. into the main
    C. Small diameter (nominally 2 in. or less) ½ diameter into the main
    D. Small diameter (nominally 2 in. or less) protruding 1 in. into the main
    E. Other (describe the limits)

Section 1.1.4.9: Can the method pass inline obstructions?
- Butterfly valves
    A. Yes, partially opened
    B. Yes, fully opened
    C. No
- Smooth bends:
    A. Yes, sharp or miter bends
    B. Yes, long smooth bends with bend radius > 3 diameters
    C. No

Section 1.1.4.10: Do pipe obstructions need to be known prior to inspection?
    A. No
    B. (Intentionally blank)
    C. Yes

The answers for the 10 questions above should be noted in the inline applicability grade card shown in Table 4-1. The scale can then be used to determine the implementation factor before continuing on to Protocol 2.

Table 4-1. Inline Applicability Grade Card

1.1.4.1	1.1.4.2	1.1.4.3	1.1.4.4	1.1.4.5	1.1.4.6	1.1.4.7	1.1.4.8	1.1.4.9	1.1.4.10

Scale:
Implementation factor Easy: Mostly As and no Cs (or worse)
Implementation factor Moderate: Mostly Bs with a few As or Cs (or worse)
Implementation factor Difficult: Mostly Bs and Cs (or worse)

Section 1.2: External Inspection

Section 1.2.1: External Condition. The external inspection category begins by determining the external condition required for the technology to be used, determined by the flowchart in Figure 4-6.

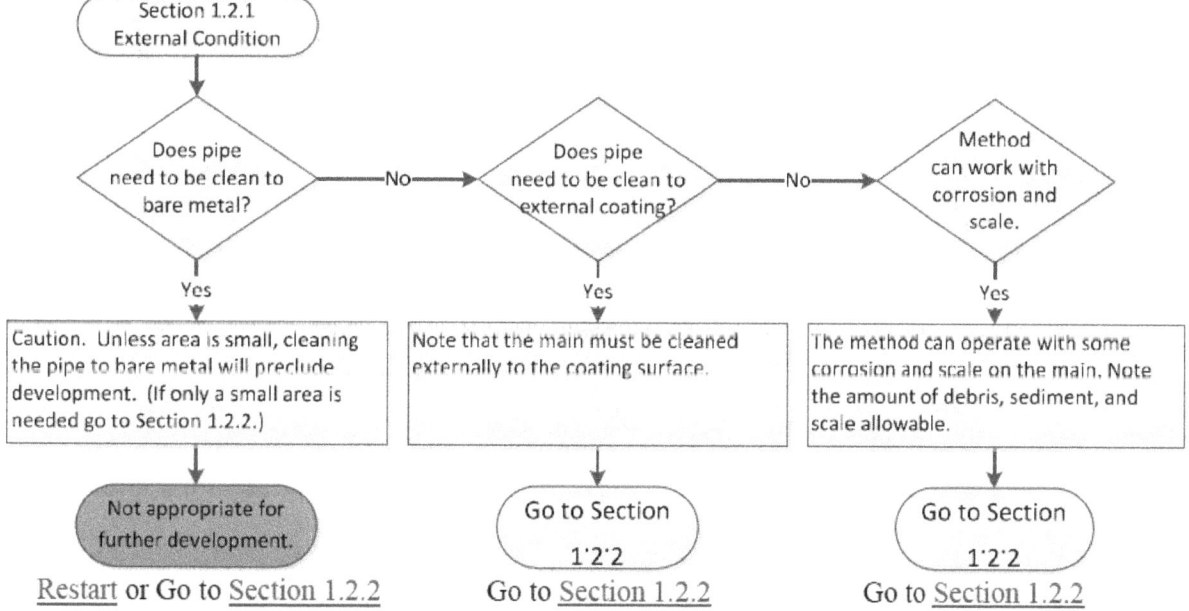

Figure 4-6. Required External Condition

Section 1.2.2: Excavation Requirements. The excavation requirements that govern the use of a structural inspection technology are described on the flowchart in Figure 4-7.

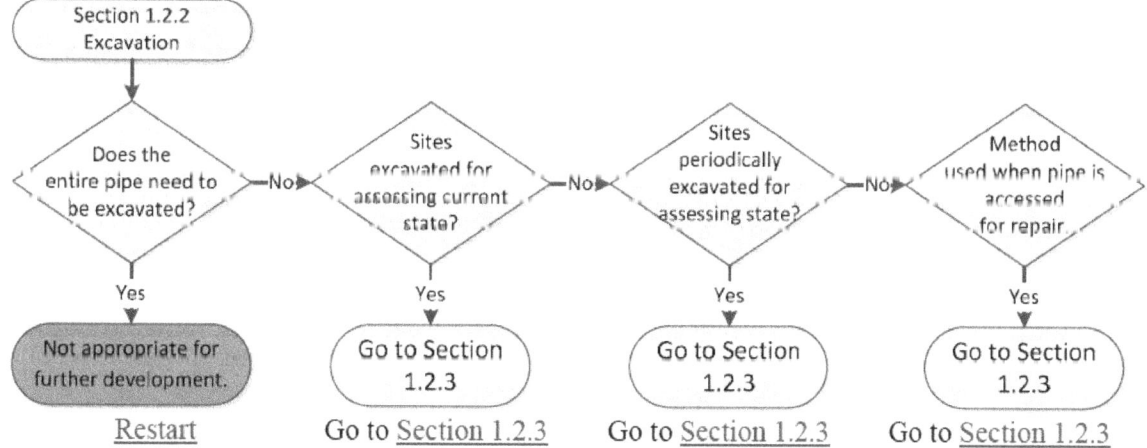

Figure 4-7. Excavation Requirements

29

Section 1.2.3: Number of Sites. If the method requires excavation, the requirements should be noted below before going to Protocol 2.

Section 1.2.3.1: How many sites per distance are needed?
> A. 1 per 1,000 ft
> B. 1 per 500 ft
> C. 1 per 200 ft
> D. 1 per 100 ft

Section 1.2.3.2: What best describes the type of excavation?
> A. Key hole
> B. Crown of pipe, over ____ ft
> C. Top half of pipe, over ____ ft
> B. Full pipe circumference, over ____ ft

Section 1.2.3.3: Do inspectors need to go into the trench?
> A. No
> C. Yes

Section 1.2.3.4: What best describes how long the sites must be open?
> A. 1 excavation at a time, open less than: (Circle one) hour, ½ day, day, 2 days, week
> B. 2 excavations at a time, open less than: (Circle one) hour, ½ day, day, 2 days, week
> C. ___ (no.) excavations at a time, open less than: (Circle one) hour, ½ day, day, 2 days, week

Section 1.2.3.5: Is traffic permitted in lanes next to excavation?
> A. Yes
> B. No

Section 1.2.3.6: Once sensors are installed, can excavations be plated and traffic maintained while inspection is being conducted remotely?
> A. Yes
> B. No

Section 1.2.3.7: Any other issues or features of the excavation?

The answers for the seven questions above should be noted in the external applicability grade card shown in Table 4-2. The scale can then be used to determine the implementation factor before continuing on to Protocol 2.

Table 4-2. External Applicability Grade Card

1.2.3.1	1.2.3.2	1.2.3.3	1.2.3.4	1.2.3.5	1.2.3.6

<u>Scale:</u>
Implementation factor Easy: Mostly As and no Cs (or worse)
Implementation factor Moderate: Mostly Bs with a few As or Cs (or worse)
Implementation factor Difficult: Mostly Bs and Cs (or worse)

Section 1.3: Above Ground Inspection

Section 1.3.1: Implementation Questions. The above ground inspection category has fewer implementation issues and hence fewer questions, as shown in Figure 4-8. Few cast iron pipes have above ground electrical connections and continuous electrical conductivity. Modifying a pipe for continuous conductivity and electrical connections would typically require excavation making the inspection an external inspection.

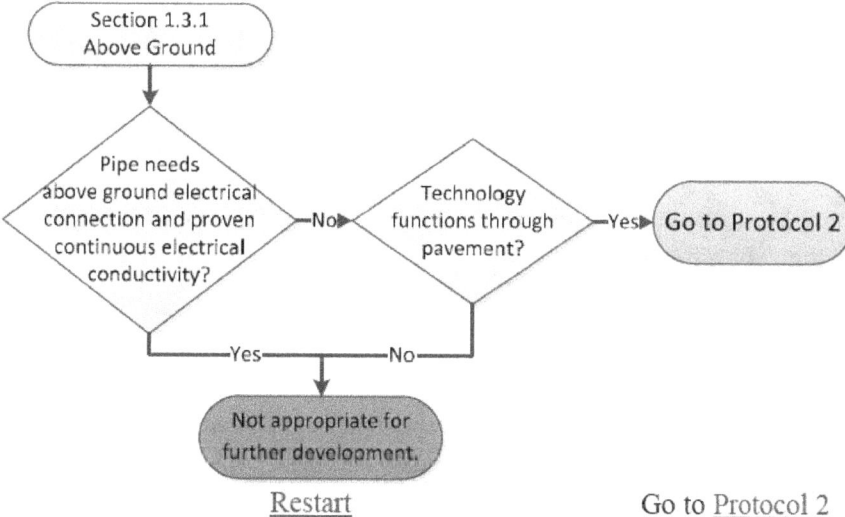

Figure 4-8. Above Ground Implementation

Section 1.4: From-the-Air Inspection of Buried Pipe

Section 1.4.1: Implementation Questions. The from the air inspection of buried pipe category has few implementation issues as shown in the flowchart in Figure 4-9.

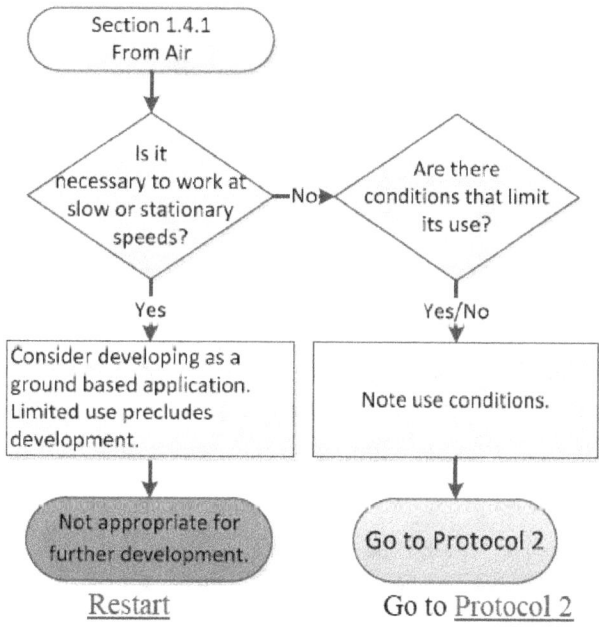

Figure 4-9. From the Air Implementation

Section 1.5: Combination Inspection

The combined inspection category includes technologies that may be evaluated multiple ways. An example of a combination technology is one that has a simple non-intrusive transmitting unit in the pipe and a robust detection system above ground. The transmitter could output acoustic or electromagnetic energy that propagates through the pipe, soil, and pavement, and could be free swimming or on a tether. This method would span both the inline and above ground categories, with the higher resolution generally provided with an inline system and the simpler implementation of an above ground system. The first category, which the technology fits in, should be chosen before repeating the process for any other applicable technologies.

Internal Inspection	**External Inspection**	**Above Ground**	**From the Air**
Go to Section 1.1	Go to Section 1.2	Go to Section 1.3	Go to Section 1.4

4.2   Secondary Screening Protocol

Protocol 2 is used to determine if the structural inspection technology is capable of detecting key distress indicators of large diameter cast iron water mains.

Section 2.0: Types of Defect Detection

The groups of key distress indicators are found in Figure 4-10 and the appropriate hyperlinks lead to the group sections.

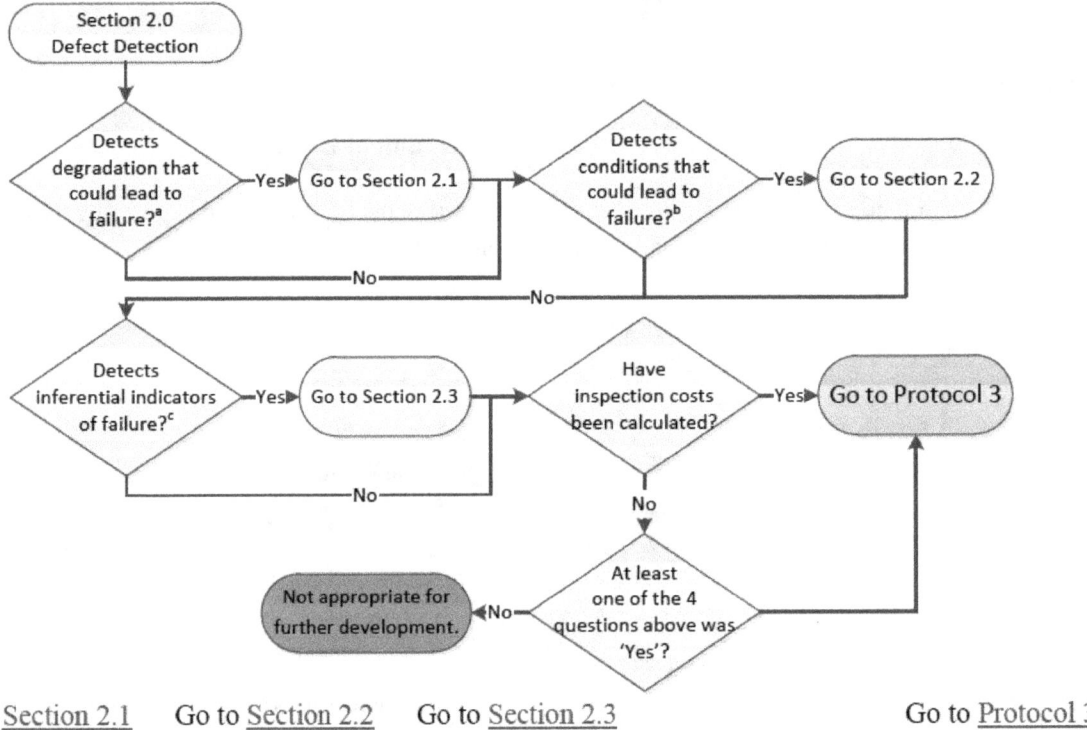

Go to Section 2.1   Go to Section 2.2   Go to Section 2.3   Go to Protocol 3

Figure 4-10. Defect Detection Categories

Note:
(a) Degradation includes corrosion, graphitization, cracks, or leakage.
(b) Conditions leading to degradation or failure include angled joints, and coating and lining defects.
(c) Inferential indicators of potential failure include pipe vintage, pressure variation, location and soil issues, and cathodic protection.

32

Section 2.1: Degradation Leading to a Failure

The types of degradation that could lead to a failure can be broken down into three areas (i.e., corrosion, graphitization, or cracks in the barrel or bell, or leaks). The ability of the technology to detect one or more of these types of degradation can be systematically characterized by completing the flowchart in Figure 4-11.

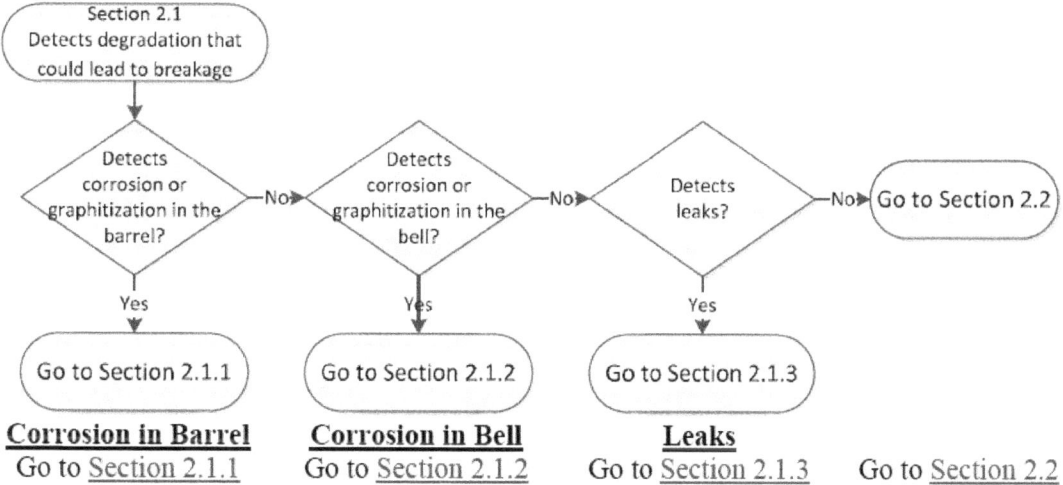

Figure 4-11. Degradation Defect Categories

Section 2.1.1: Corrosion in the Pipe Barrel. The flowchart in Figure 4-12 collects data about corrosion detection in the pipe barrel.

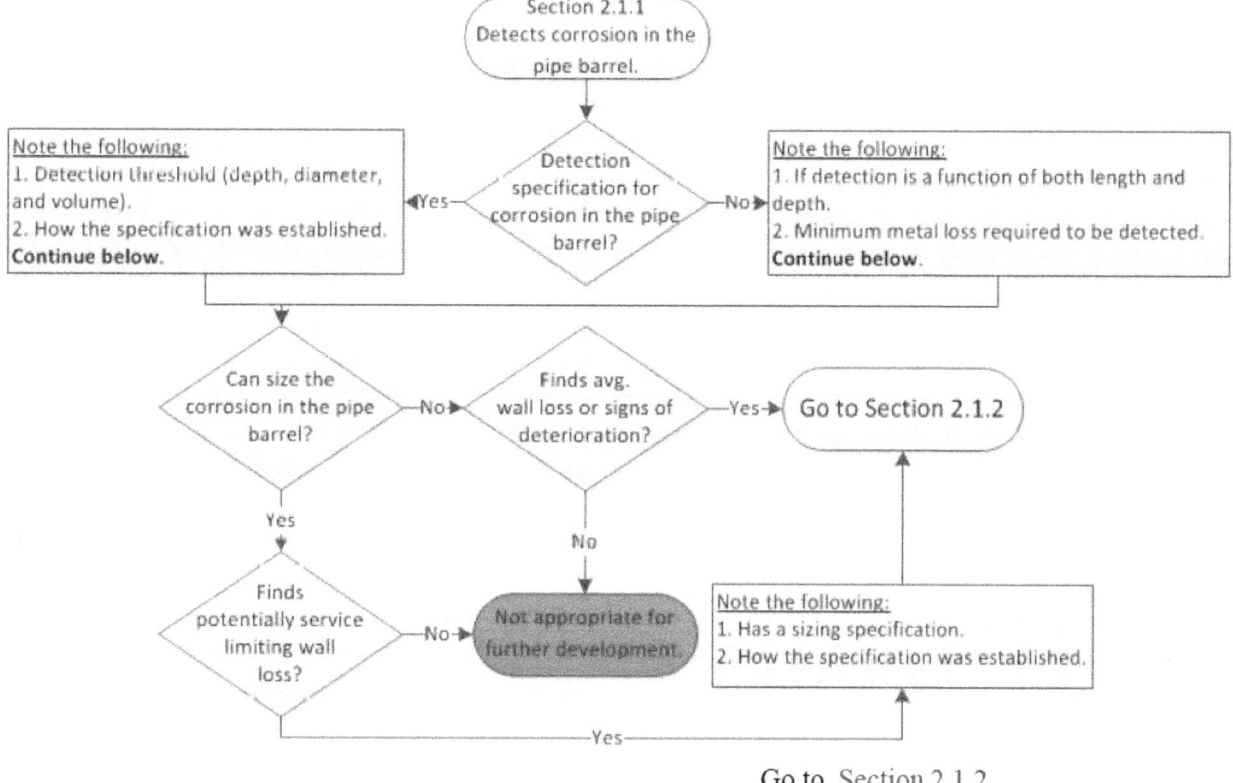

Figure 4-12. Corrosion Detection in the Pipe Barrel

Section 2.1.2: Corrosion in the Bell. The flowchart in Figure 4-13 collects data about corrosion detection in the bell.

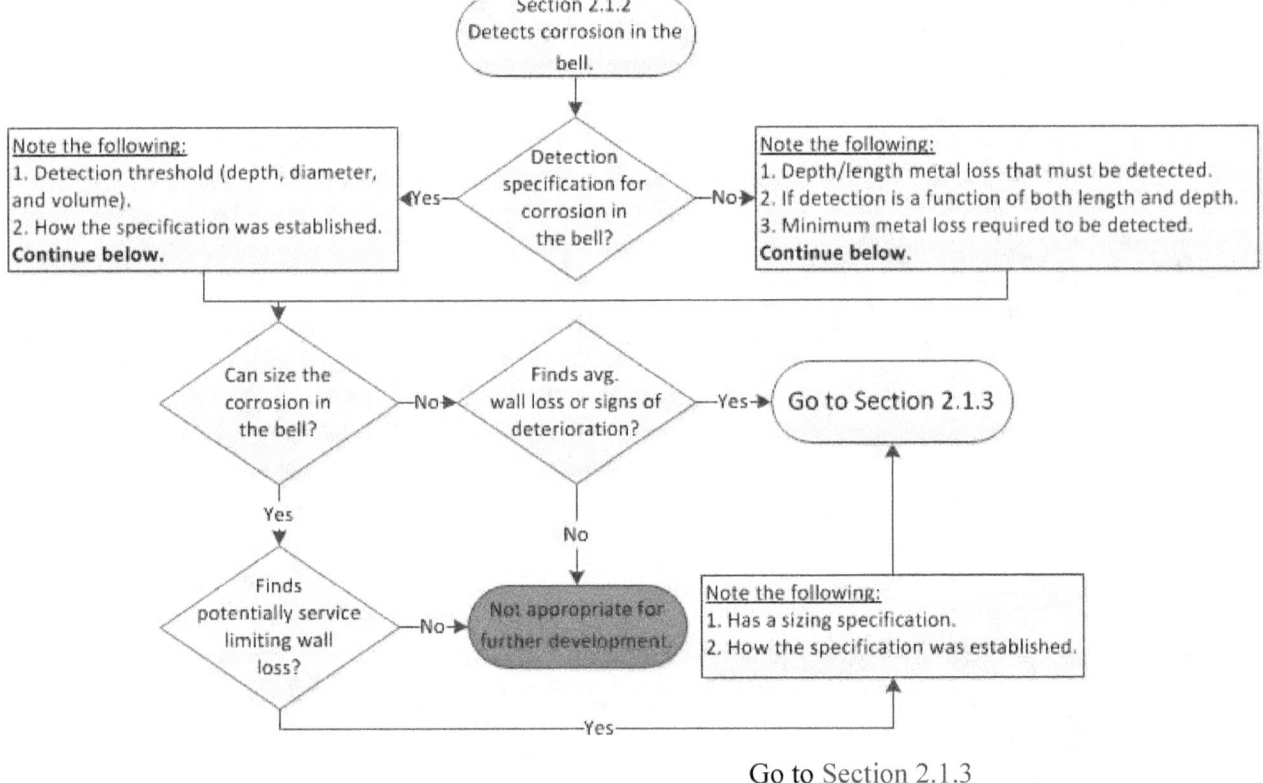

Figure 4-13. Corrosion Detection in the Bell

Section 2.1.3: Leaks. The flowchart in Figure 4-14 collects data about leak detection.

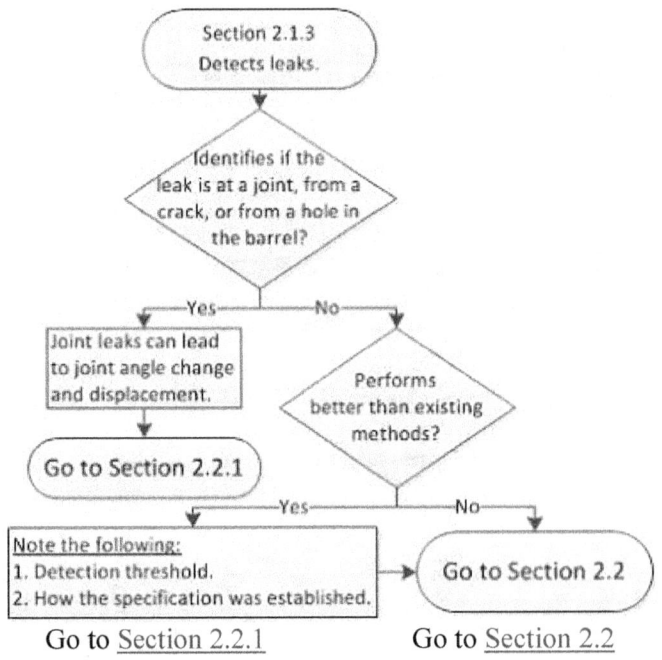

Figure 4-14. Leak Detection

Section 2.2: Conditions Leading to Degradation

The types of conditions that could lead to degradation or failure can be broken down into three areas (i.e., pipe angle, internal linings, and external coatings) determined by the flowchart in Figure 4-15.

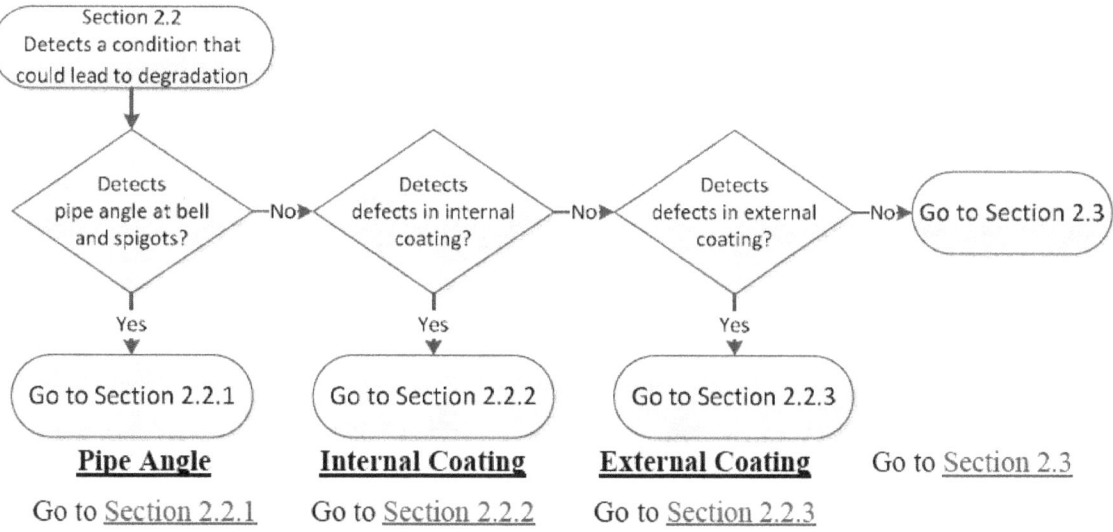

Figure 4-15. Conditions Leading to Degradation or Failure

Section 2.2.1: Angle between Pipe Lengths at the Bell and Spigot. The flowchart in Figure 4-16 collects data about the pipe angle at the bell and spigot.

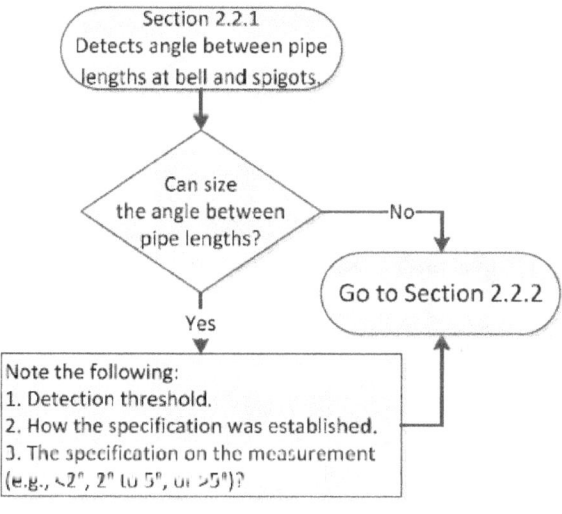

Go to Section 2.2.2

Figure 4-16. Pipe Angle Between the Bell and Spigot

Section 2.2.2: Defects in the Internal Coating. The flowchart in Figure 4-17 collects data about defect in the internal coating.

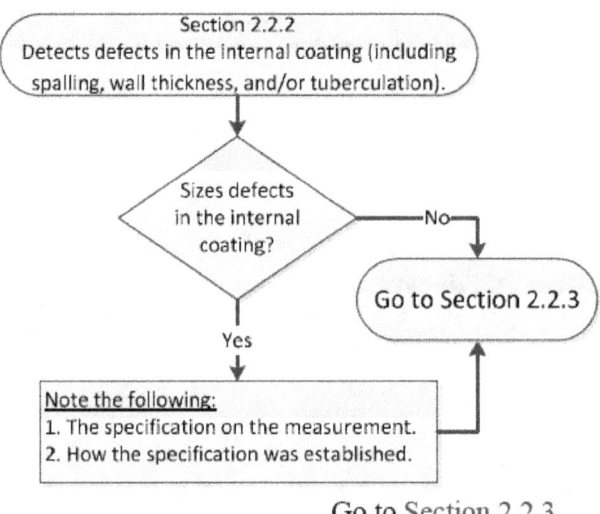

Go to Section 2.2.3

Figure 4-17. Defects in the Internal Coating

Section 2.2.3: Defects in the External Coating. The flowchart in Figure 4-18 collects data about defects in the external coating.

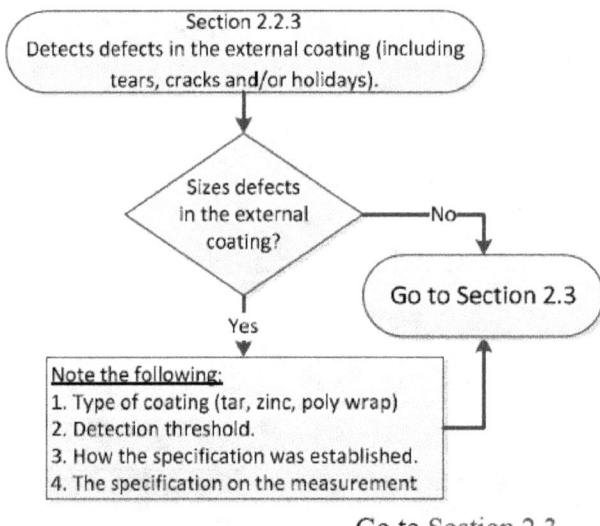

Go to Section 2.3

Figure 4-18. Defects in the External Coating

Section 2.3: Inferential Indicators of Failure

The inferential indicators of pipes with a higher probability of failure can be broken down into five areas (i.e., pipe vintage, pressure variations, location issues, soils issues and cathodic protection) determined by the flowchart in Figure 4-19. Technologies that detect water quality issues are not considered structural inspection technologies.

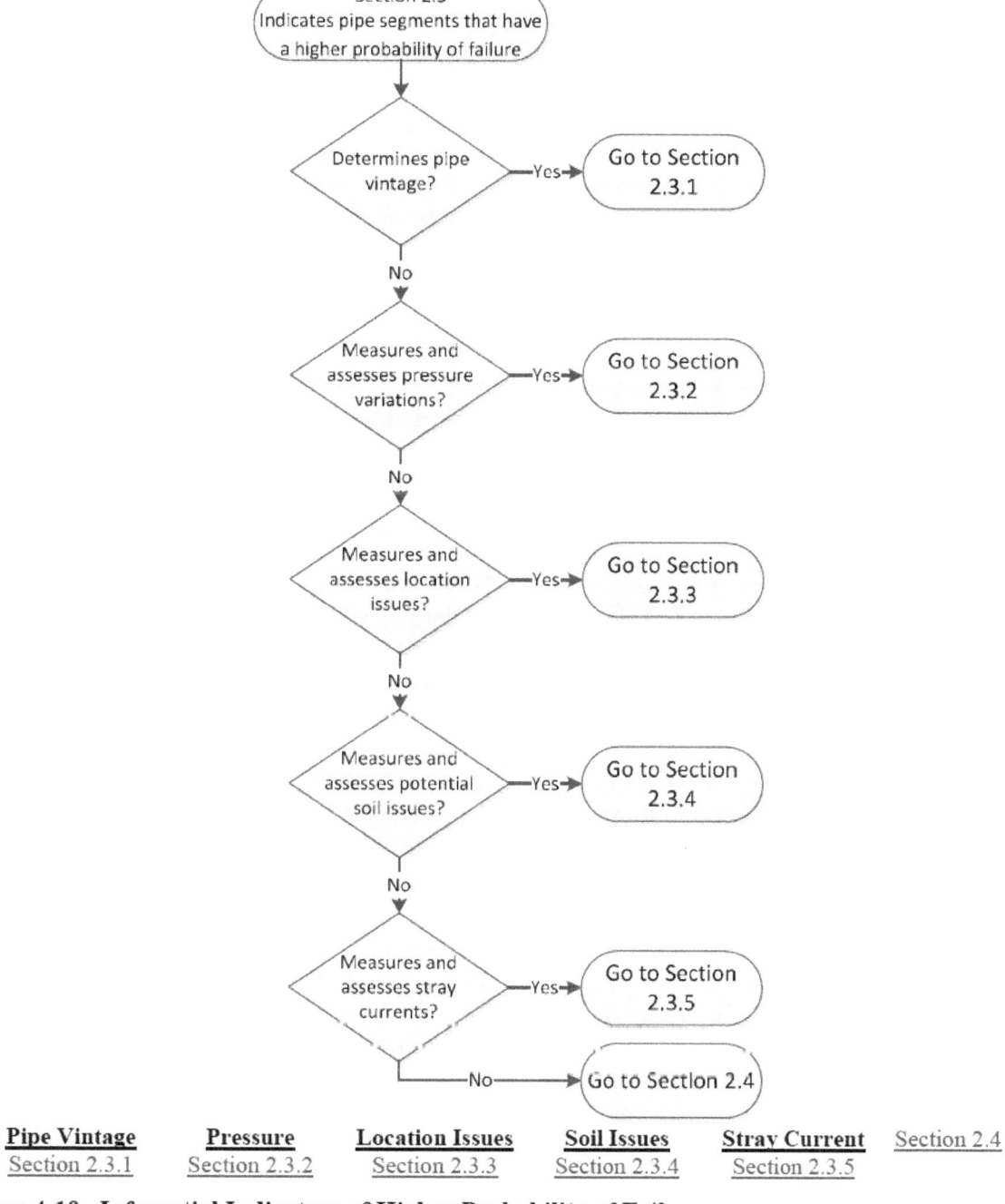

Figure 4-19. Inferential Indicators of Higher Probability of Failure

Section 2.3.1: Pipe Vintage. The flowchart in Figure 4-20 is used to collect data about pipe vintage, nominal wall thickness, pipe properties, or other baseline values that would help determine operational factors.

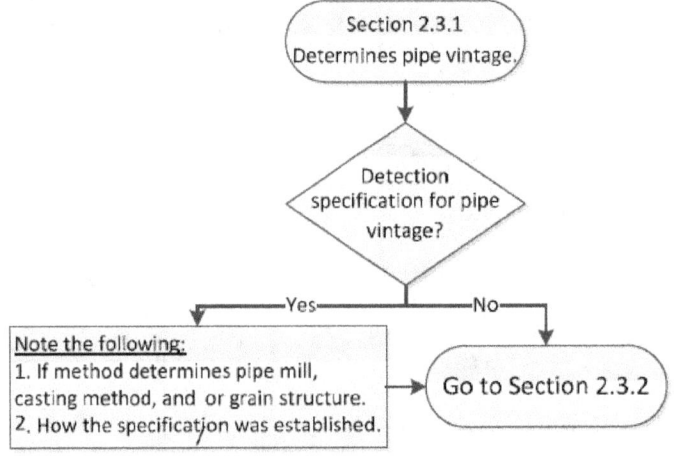

Go to Section 2.3.2

Figure 4-20. Determines Pipe Vintage

Section 2.3.2: Pressure Variations. The flowchart in Figure 4-21 is used to collect data about pressure change.

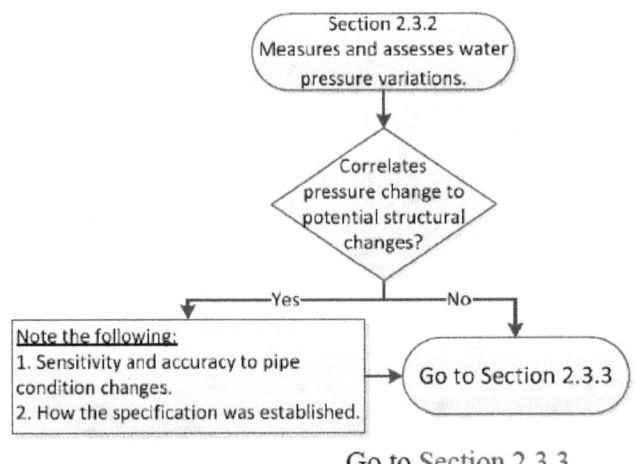

Go to Section 2.3.3

Figure 4-21. Measures and Assesses Water Pressure Variations

Section 2.3.3: Location Issues. The flowchart in Figure 4-22 is used to collect data about location issues.

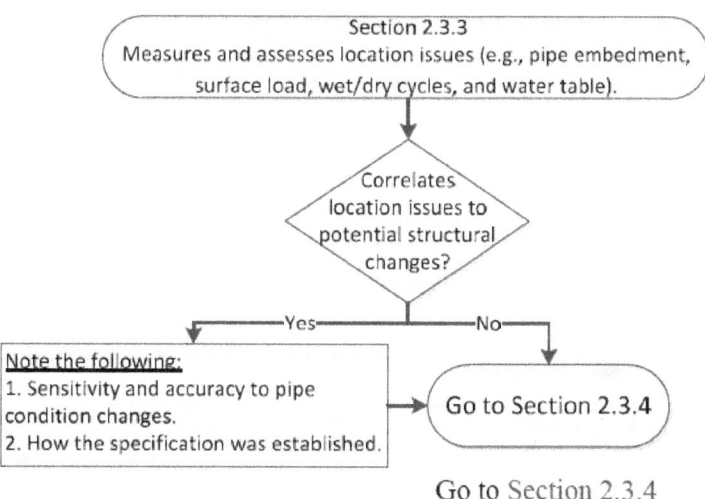

Figure 4-22. Assesses Location Issues

Section 2.3.4: Soil Issues. The flowchart in Figure 4-23 is used to collect data about soil issues.

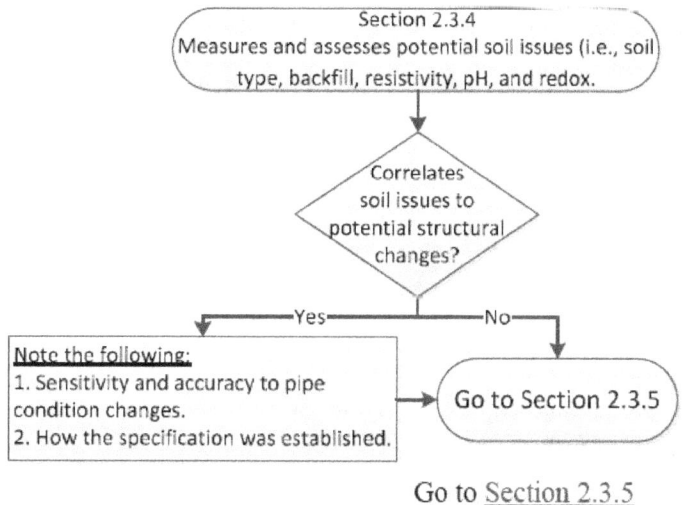

Figure 4-23. Assesses Soil Issues

Section 2.3.5: Cathodic Protection. The flowchart in Figure 4-24 is used to collect data about cathodic protection and stray currents.

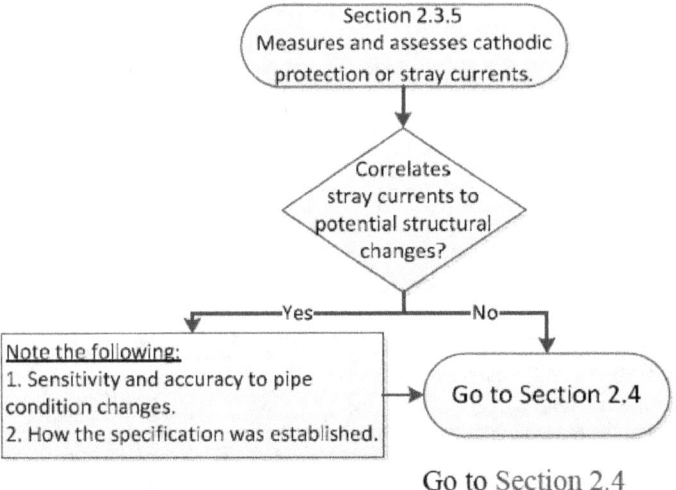

Go to Section 2.4

Figure 4-24. Assesses Cathodic Protection and Stray Currents

Section 2.4: System Cost and On-site Inspection Costs

If system and on-site inspection costs are known, the following questions can be used to determine the relative cost grade for using and developing the technology further. Note all answers and calculate the cost grade before moving to Protocol 3.

Section 2.4.1: What is the capital cost of an inspection system?
A. Less than $50,000
B. Between $50,000 and $200,000
C. Between $200,000 and $500,000
D. Over $500,000

Section 2.4.2: How much more capital is needed to complete the technology development?
A. Less than $100,000
B. Between $100,000 and $400,000
C. Between $400,000 and $1,000,000
D. Over $1,000,000

Section 2.4.3: How many technicians are needed on-site to apply the technology?
A. 1
B. 2
C. 3
D. more than 3

Section 2.4.4: Transporting equipment?
A. Carry on or checked baggage
B. Shipping
C. Dedicated truck

Section 2.4.5: How many man days are needed to analyze one day of data?
A. Same day on-site or next day
B. Within 2 weeks
C. Within 1 month
D. More than 1 month

Section 2.4.6: How many feet can be inspected in one day?
A. More than 10,000 feet
B. 4,000 to 10,000 feet
C. Distance between valves, nominally 2,000 feet
D. Less than 1000 feet

Section 2.4.7: What is the estimated cost to modify the pipeline for assessment? What are the costs to return the pipeline to operation?
A. 1 day for a typical contractor, fittings around $100
B. 2 - 3 days for a typical contractor, fittings around $500
C. 1 - 2 weeks for a typical contractor, fittings around $1,000
D. More than 2 weeks, fittings more than $1,000

Section 2.4.8: What is the basis for these costs?

The answers for the eight questions above should be noted in the cost grade card shown in Table 4-3. The scale can then be used to determine the cost factor before continuing on to Protocol 3.

Table 4-3. Cost Grade Card

2.4.1	2.4.2	2.4.3	2.4.4	2.4.5	2.4.6	2.4.7

<u>Scale:</u>
Implementation cost Low: Mostly As and no Cs (or worse)
Implementation cost Medium: Mostly Bs with a few As or Cs (or worse)
Implementation cost High: Mostly Bs and Cs (or worse)

4.3   Tertiary Screening Protocol

Protocol 3 is used to determine how the new structural inspection technology compares to existing technologies and whether the technology has the potential for further development for application to water mains.

Section 3.0: Types of Defect Detection

Protocol 3 begins by selecting the types of defects the technology can detect based on the answers from Protocol 2 shown in the flowchart below (Figure 4-25).

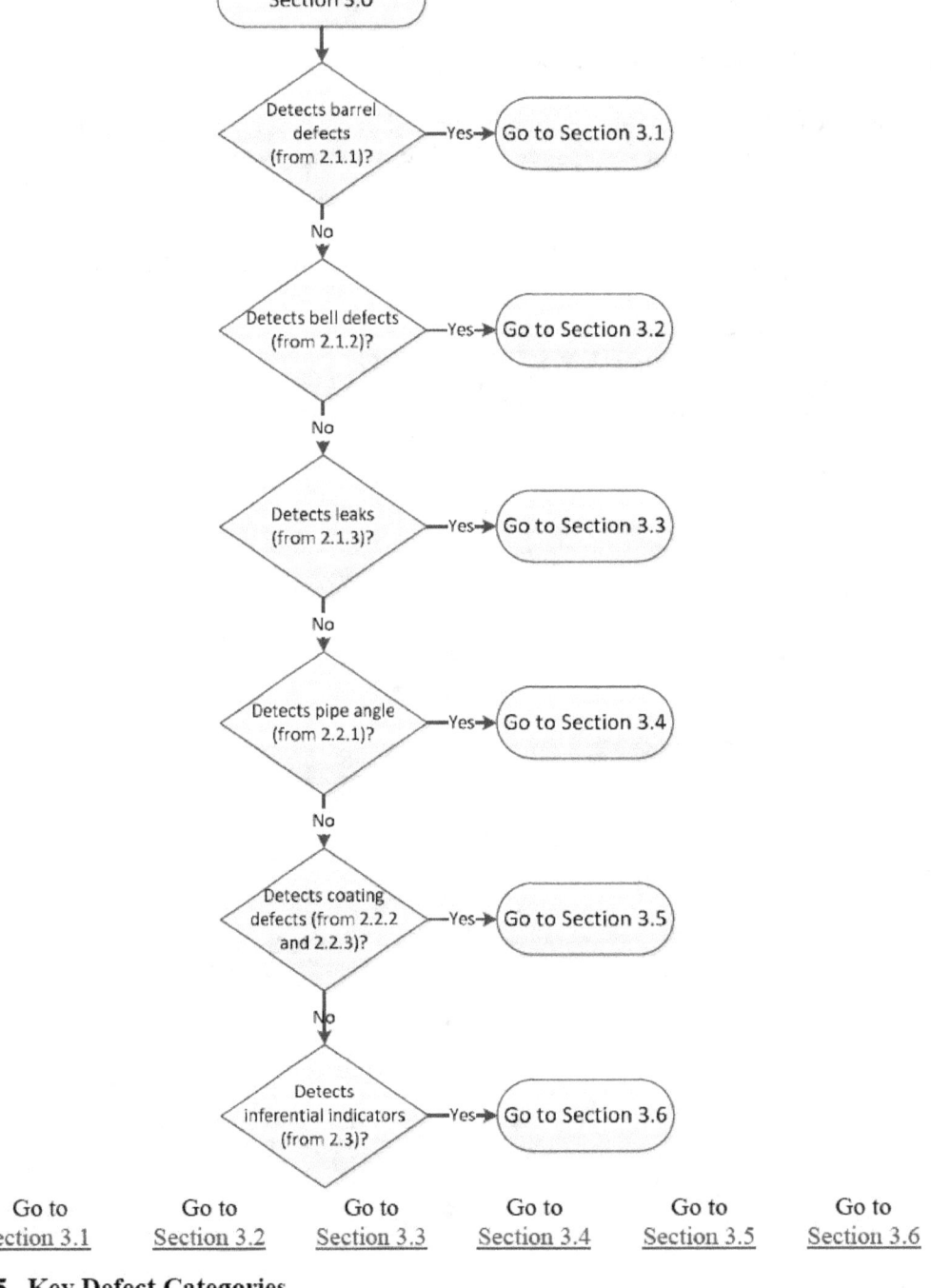

Figure 4-25. Key Defect Categories

42

Section 3.1: Detects Barrel Corrosion

The types of inspections for detecting barrel corrosion are outlined by the flowchart in Figure 4-26.

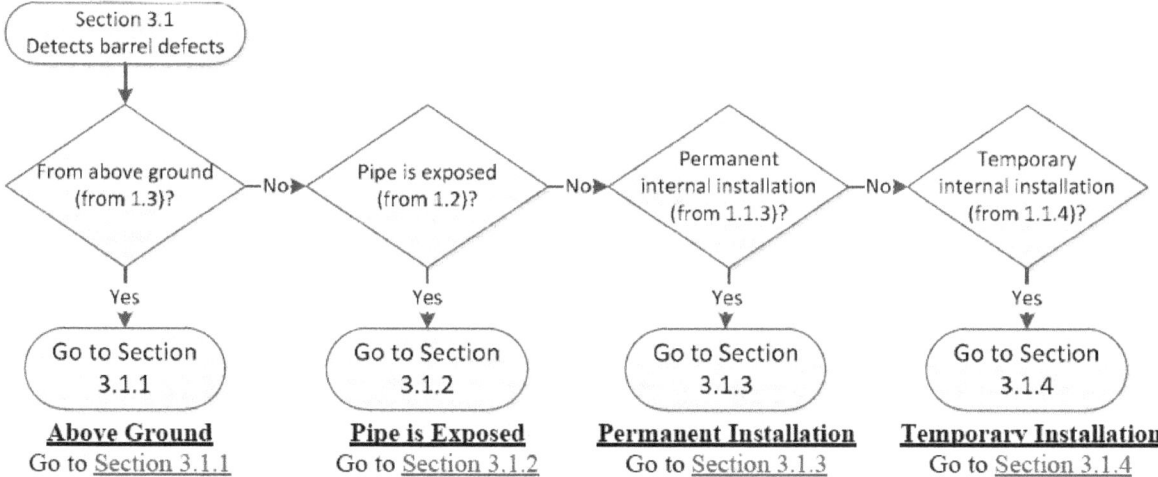

Figure 4-26. Detecting Barrel Corrosion

Section 3.1.1: Barrel Corrosion, Above Ground. The flowchart in Figure 4-27 is used to determine which approaches for detecting barrel corrosion from above ground have the potential for further development. All cost categories are determined from Section 2.4 (see Table 4-3).

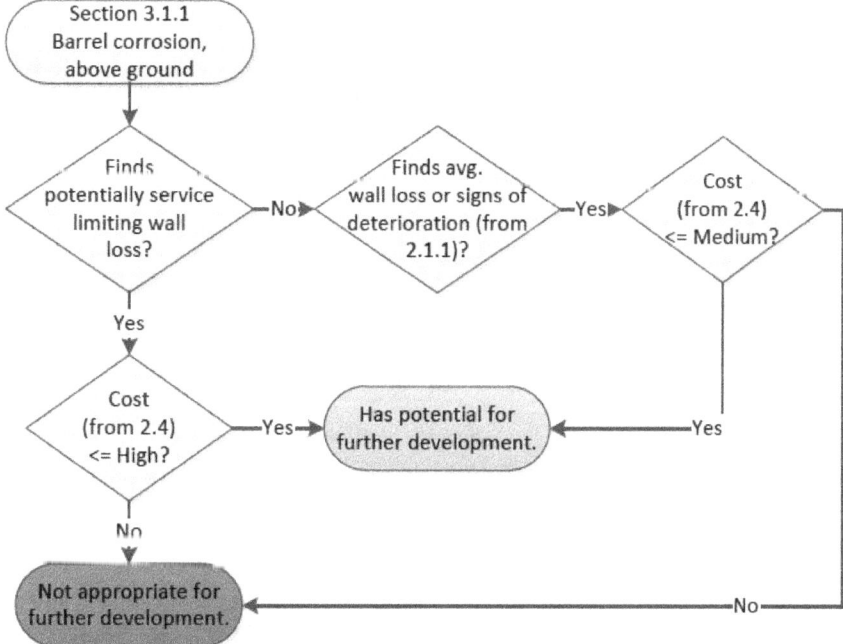

Figure 4-27. Potential for Detecting Barrel Corrosion from Above Ground

Section 3.1.2: Barrel Corrosion, Pipe is Exposed. The flowchart in Figure 4-28 is used to determine which approaches for detecting barrel corrosion on an exposed pipe have the potential for further development.

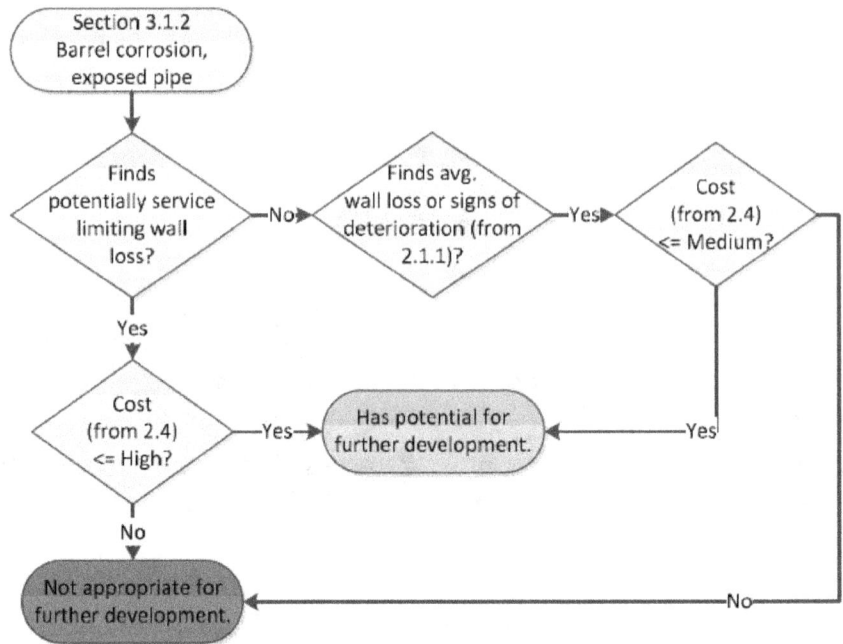

Figure 4-28. Potential for Detecting Barrel Corrosion on an Exposed Pipe

Section 3.1.3: Barrel Corrosion, Permanent Internal Installation. The flowchart in Figure 4-29 is used to determine which approaches for detecting barrel corrosion using an internal device permanently installed in the main have the potential for further development.

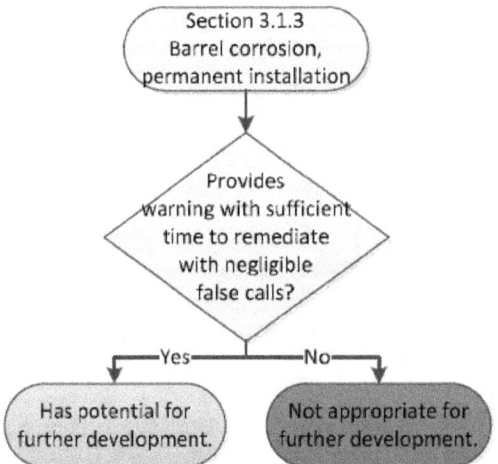

Figure 4-29. Potential for Detecting Barrel Corrosion with a Permanently Installed Internal Device

44

Section 3.1.4: Barrel Corrosion, Temporary Internal Installation. The flowchart in Figure 4-30 is used to determine which approaches for detecting barrel corrosion using an internal device temporarily installed in the main have the potential for further development.

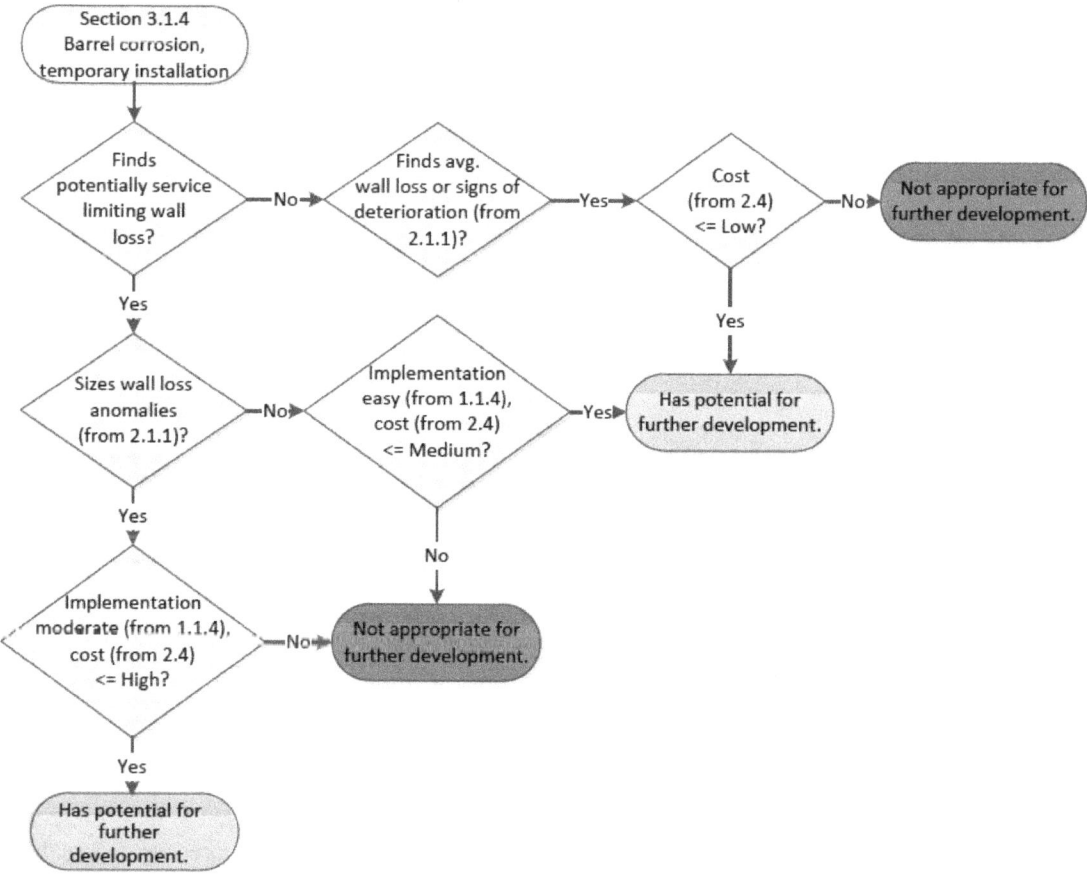

Figure 4-30. Potential for Detecting Barrel Corrosion with a Temporarily Installed Internal Device

Section 3.2: Detects Bell Corrosion and Cracks

The types of inspections for detecting bell defects are outlined by the flowchart in Figure 4-31.

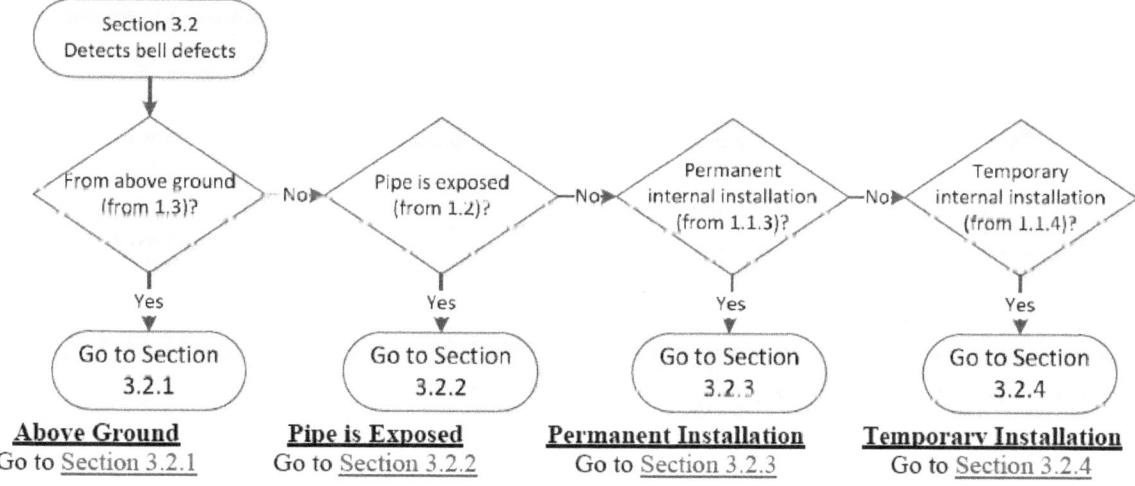

Above Ground
Go to Section 3.2.1

Pipe is Exposed
Go to Section 3.2.2

Permanent Installation
Go to Section 3.2.3

Temporary Installation
Go to Section 3.2.4

Figure 4-31. Detecting Bell Corrosion and Cracks

Section 3.2.1: Bell Defects, Above Ground Techniques. The flowchart in Figure 4-32 is used to determine which approaches for detecting bell corrosion and cracks from above ground have the potential for further development.

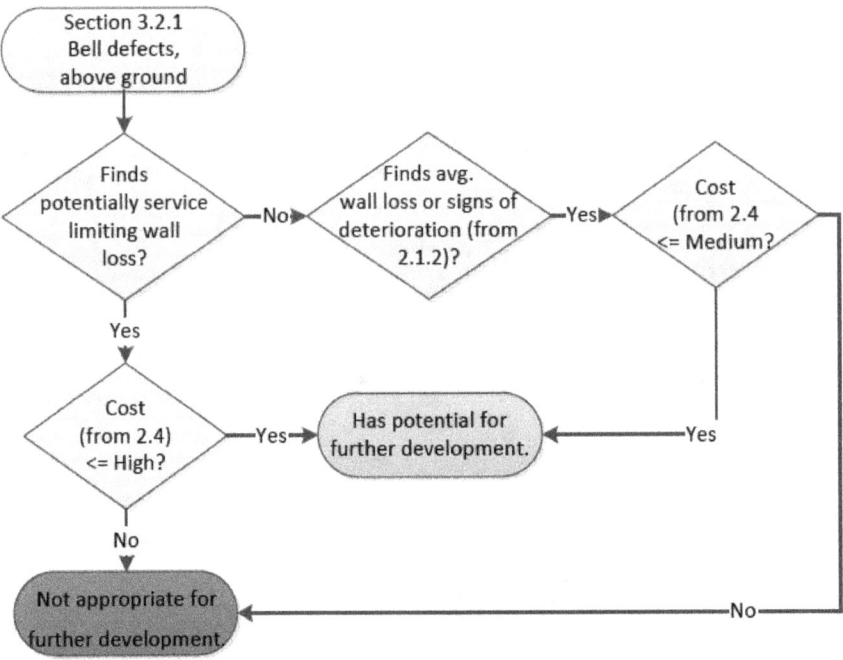

Figure 4-32. Potential for Detecting Bell Defects from Above Ground

Section 3.2.2: Bell Defects, Pipe is Exposed. The flowchart in Figure 4-33 is used to determine which approaches for detecting bell corrosion and cracks on an exposed pipe have the potential for further development.

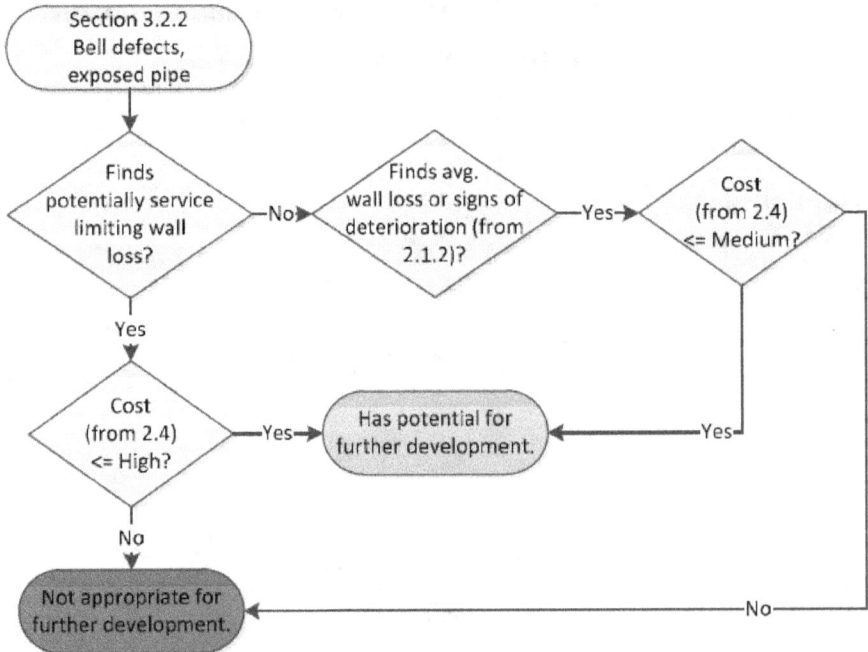

Figure 4-33. Potential for Detecting Bell Defects on an Exposed Pipe

Section 3.2.3: Bell Defects, Permanent Internal Installation. The flowchart in Figure 4-34 is used to determine which approaches for detecting bell corrosion and cracks using an internal device permanently installed in the main have the potential for further development.

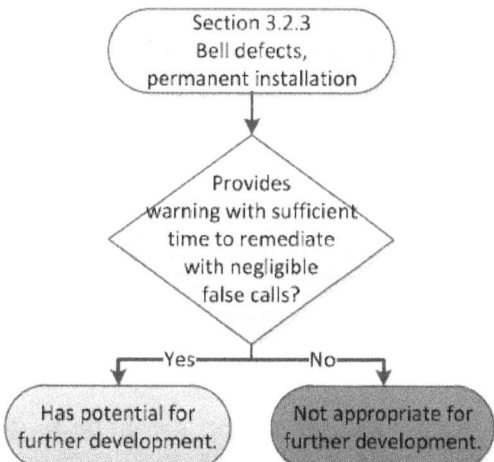

Figure 4-34. Potential for Detecting Bell Defects with a Permanently Installed Internal Device

Section 3.2.4: Bell Defects, Temporary Internal Installation. The flowchart in Figure 4-35 is used to determine which approaches for detecting bell corrosion and cracks using an internal device temporarily installed in the main have the potential for further development.

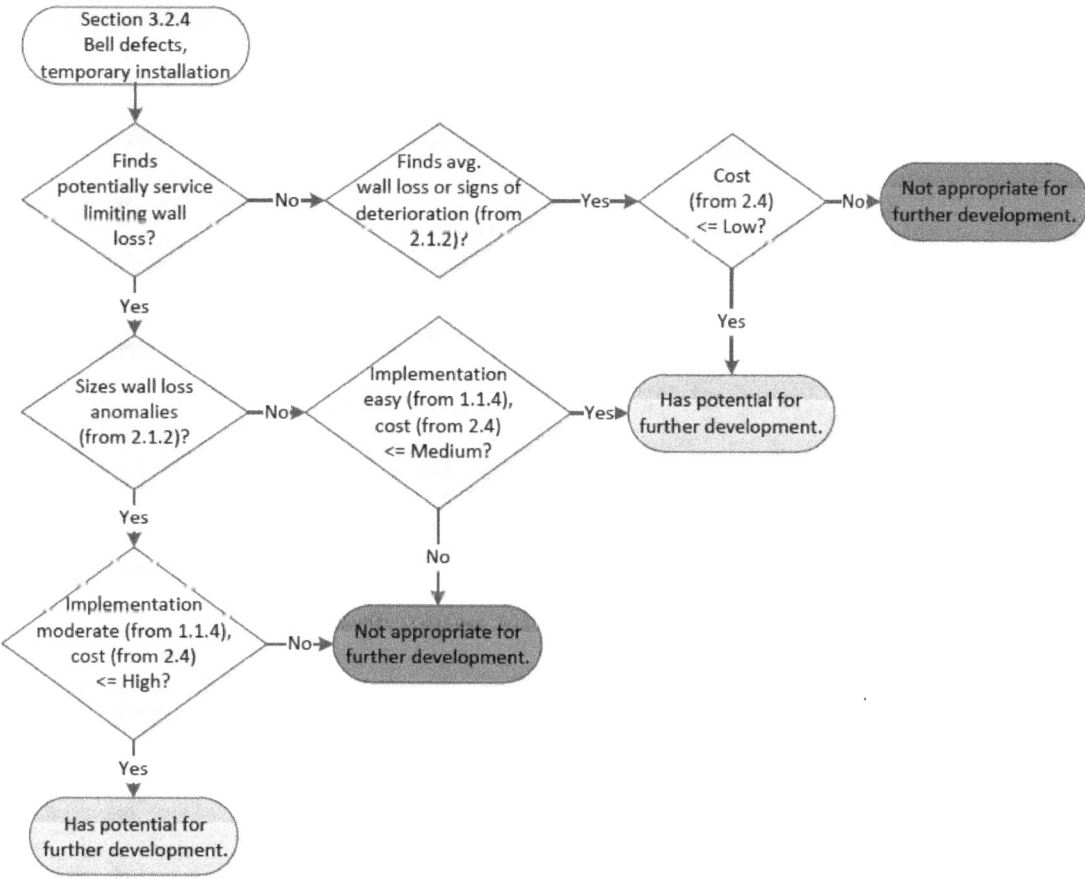

Figure 4-35. Potential for Detecting Bell Defects with a Temporarily Installed Internal Device

47

Section 3.3: Detects Leaks

The types of inspections for detecting leaks are outlined by the flowchart in Figure 4-36.

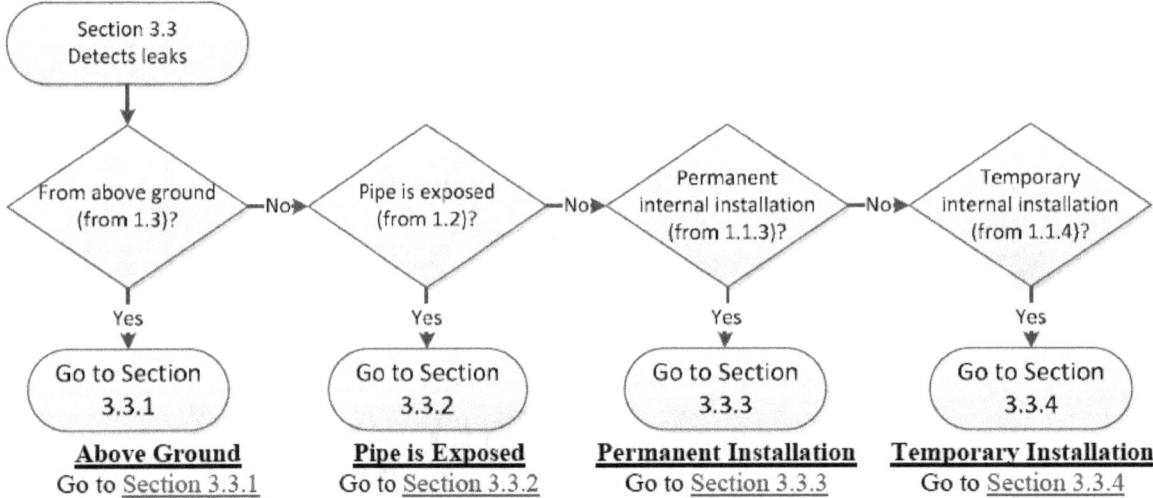

Figure 4-36. Detecting Leaks

Section 3.3.1: Leaks, Above Ground. The flowchart in Figure 4-37 is used to determine which approaches for detecting leaks from above ground have the potential for further development.

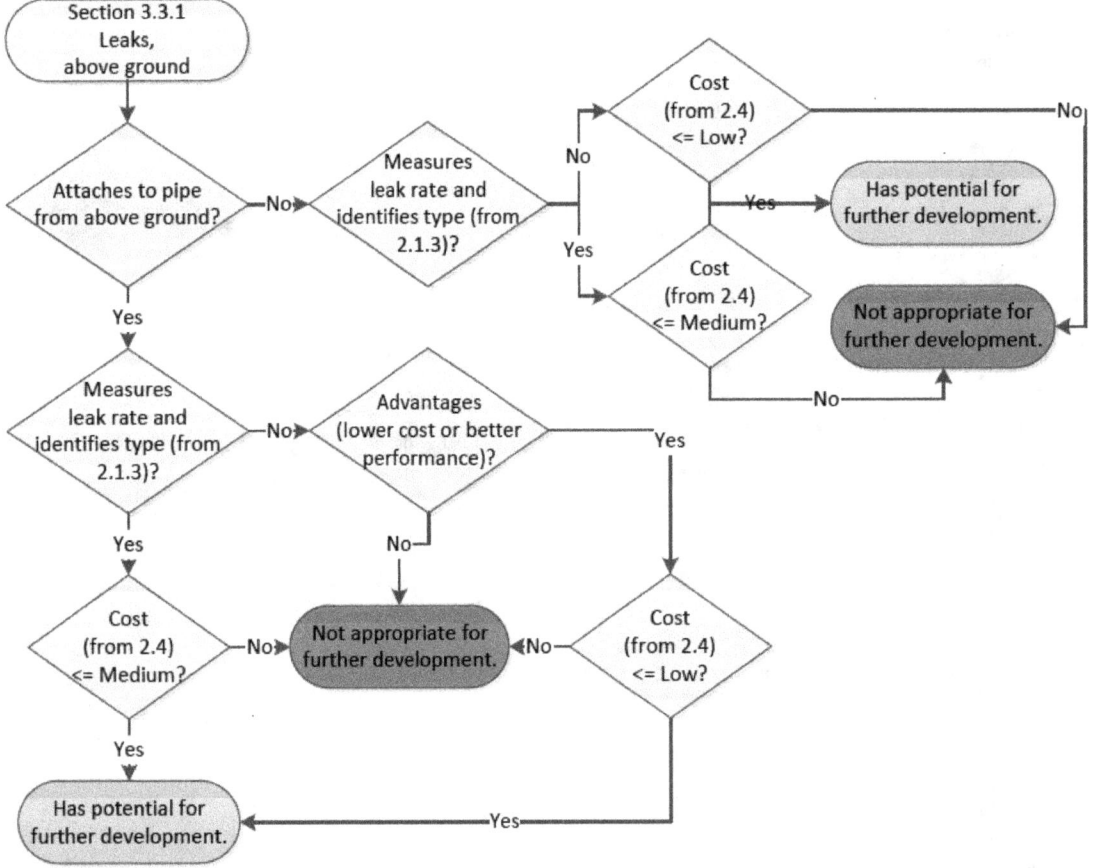

Figure 4-37. Potential for Detecting Leaks from Above Ground

Section 3.3.2: Leaks, Pipe is Exposed. The flowchart in Figure 4-38 is used to determine which approaches for detecting leaks on an exposed pipe have the potential for further development.

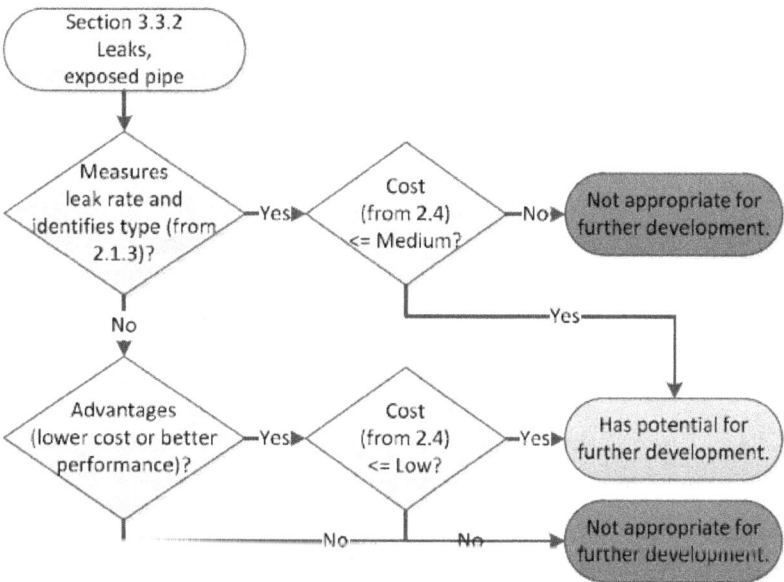

Figure 4-38. Potential for Detecting Leaks on an Exposed Pipe

Section 3.3.3: Leaks, Permanent Internal Installation. The flowchart in Figure 4-39 is used to determine which approaches for detecting leaks using an internal device permanently installed in the main have the potential for further development.

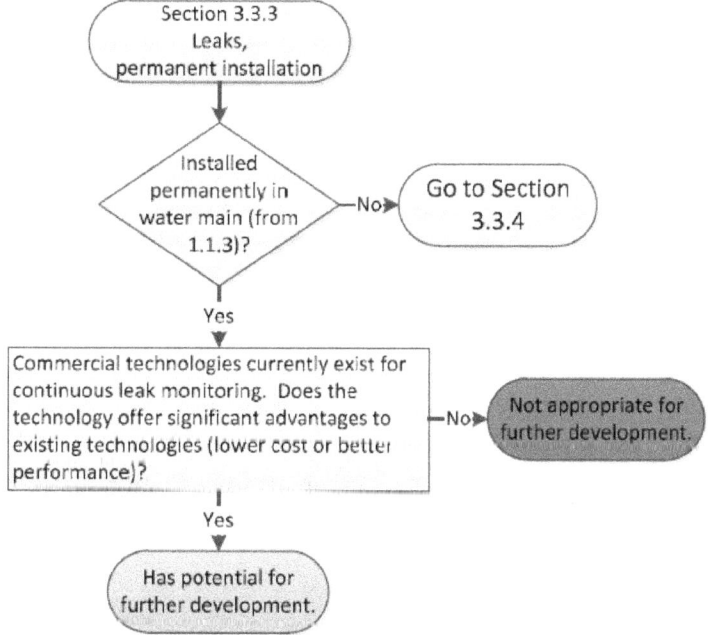

Figure 4-39. Potential for Detecting Leaks with a Permanently Installed Internal Device

Section 3.3.4: Leaks, Temporary Internal Installation. The flowchart in Figure 4-40 is used to determine which approaches for detecting leaks using an internal device temporarily installed in the main have the potential for further development.

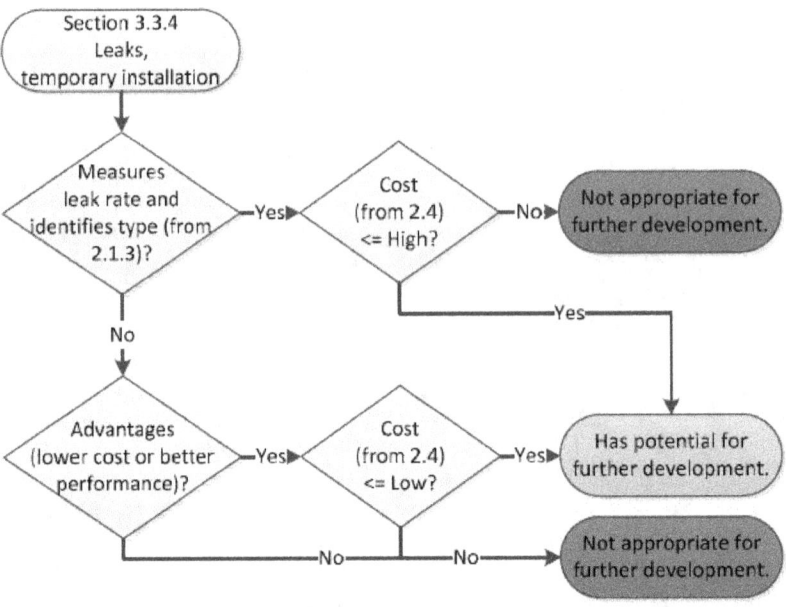

Figure 4-40. Potential for Detecting Leaks with a Temporarily Installed Internal Device

Section 3.4: Detects Pipe Angle

The types of inspections for detecting the pipe angle between bells and spigots are outlined by the flowchart in Figure 4-41.

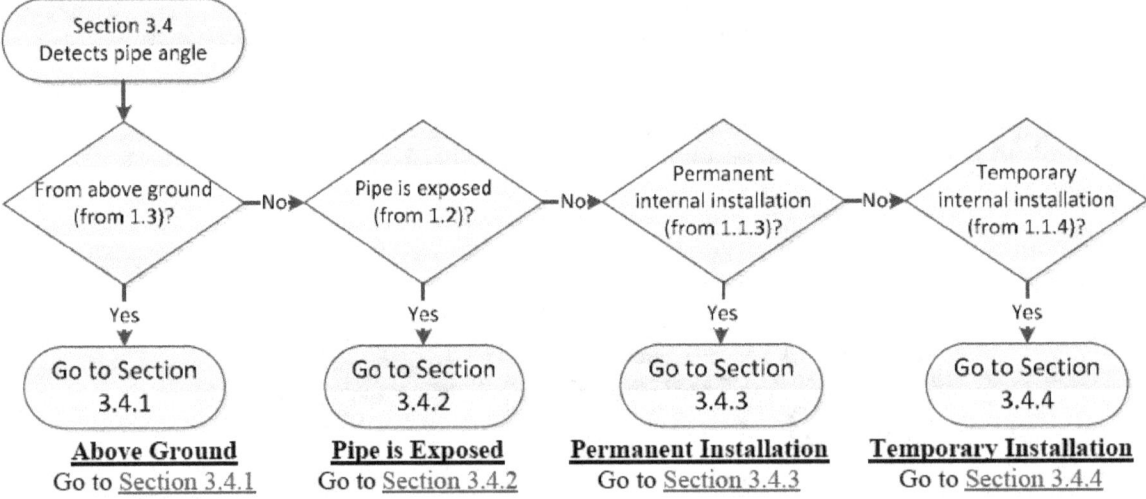

Figure 4-41. Detecting Pipe Angle

Section 3.4.1: Pipe Angle, Above Ground. The flowchart in Figure 4-42 is used to determine which approaches for detecting pipe angles from above ground have the potential for further development.

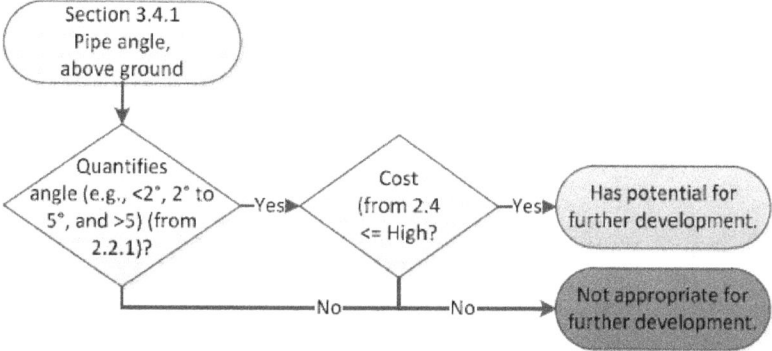

Figure 4-42. Potential for Detecting Pipe Angle from Above Ground

Section 3.4.2: Pipe Angle, Pipe is Exposed. The flowchart in Figure 4-43 is used to determine which approaches for detecting pipe angles on an exposed pipe have the potential for further development.

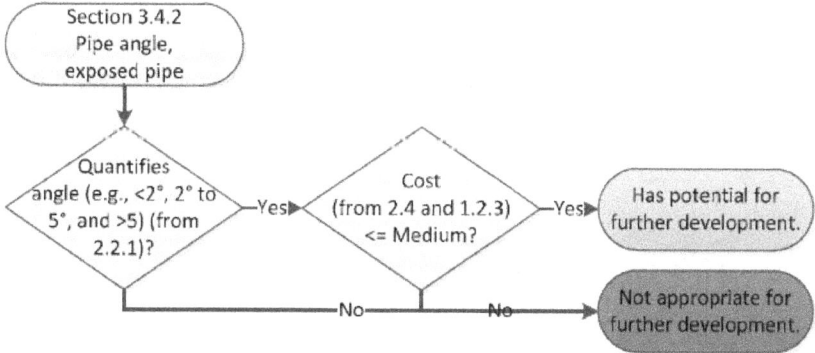

Figure 4-43. Potential for Detecting Pipe Angle on an Exposed Pipe

Section 3.4.3: Pipe Angle, Permanent Internal Installation. The flowchart in Figure 4-44 is used to determine which approaches for detecting pipe angles using an internal device permanently installed in the main have the potential for further development.

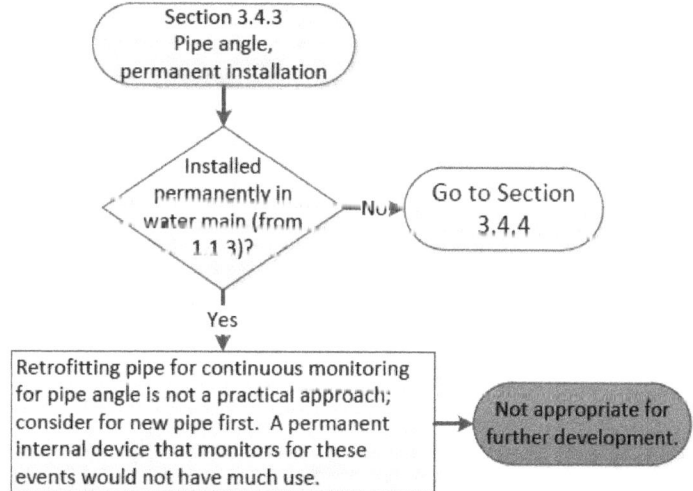

Figure 4-44. Potential for Detecting Pipe Angle with a Permanently Installed Internal Device

Section 3.4.4:   Pipe Angle, Temporary Internal Installation. The flowchart in Figure 4-45 is used to determine which approaches for detecting pipe angles using an internal device temporarily installed in the main have the potential for further development.

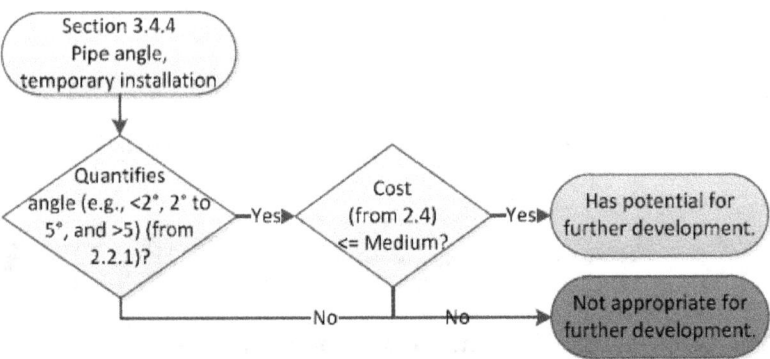

Figure 4-45. Potential for Detecting Pipe Angle with a Temporarily Installed Internal Device

Section 3.5: Detects Coating Defects

It is not likely that a water utility would only look for coating defects using a standalone system. The method may have potential for further development if the assessment method would augment another technology for minimal additional cost. The types of coatings that can be detected are outlined by the flowchart in Figure 4-46.

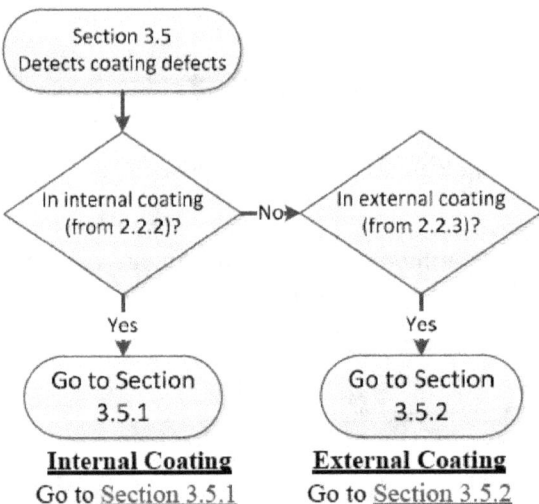

Figure 4-46. Detecting Coating Defects

Section 3.5.1: Internal Coating Defects. The flowchart in Figure 4-47 is used to determine which approaches for detecting internal coating defects have the potential for further development.

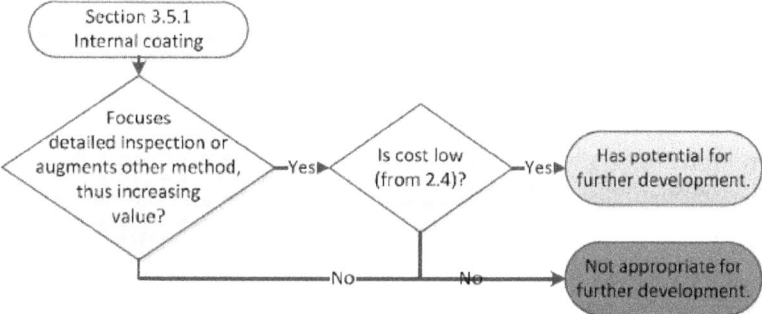

Figure 4-47. Potential for Detecting Internal Coating Defects

Section 3.5.2: External Coating Defects. The flowchart in Figure 4-48 is used to determine which approaches for detecting external coating defects have the potential for further development.

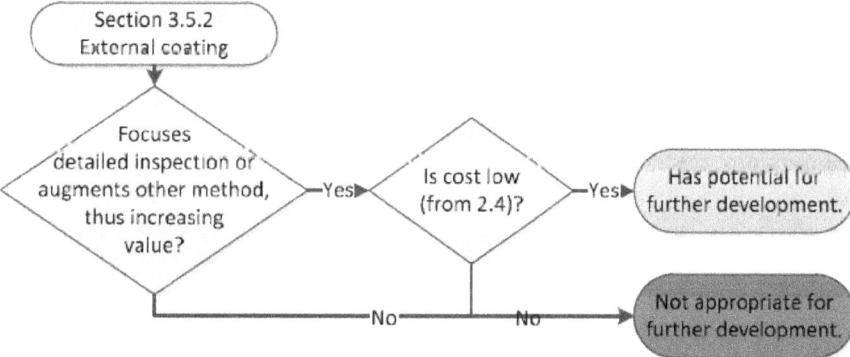

Figure 4-48. Potential for Detecting External Coating Defects

Section 3.6: Detects Inferential Indicators

It is not likely that a water utility would only look for inferential indicators using a standalone system. The method may have potential for further development if the assessment method could be used to focus detailed inspections or would augment another technology for minimal additional cost. The flowchart in Figure 4-49 is used to determine which approaches for detecting inferential indicators have the potential for further development.

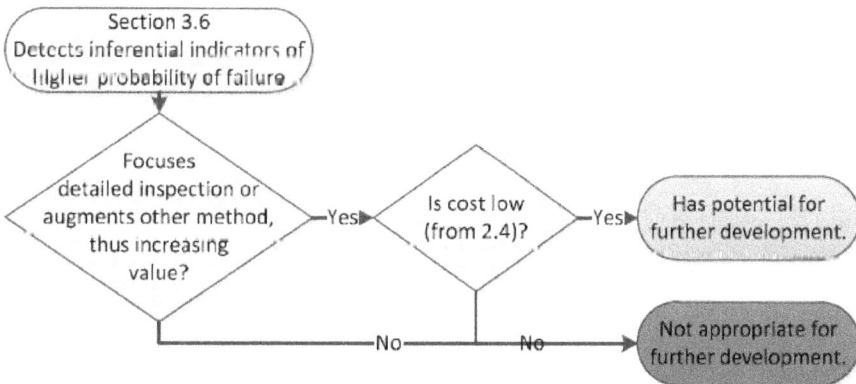

Figure 4-49. Potential for Detecting Inferential Indicators of Failure

4.4 Protocol Summary

The three protocols described herein provide agencies that could potentially fund structural inspection technology research and development with a process for determining if promising technologies have the potential for further development. The protocol first determines if a technology is applicable to large diameter cast iron water mains (i.e., Protocol 1); then it determines if the technology is applicable to detecting the key distress indicators of large diameter cast iron pipe (i.e., Protocol 2); and finally, based on the responses from Protocols 1 and 2, it determines the potential for further development (Protocol 3). The following section demonstrates the application of the protocols on eight technologies that could potentially be used for structural inspection of large diameter cast iron water mains.

5.0: APPLICATION OF PROTOCOLS

5.1   Overview

The purpose of this section is to demonstrate the use of the structural inspection technology evaluation protocols outlined in Section 4. The goal is to determine if the protocol is implementable and produces reasonable results. Four general types of both existing and emerging technologies (i.e., eight total technologies) were examined. A brief description of each technology is provided before applying the protocols.

An overview of the existing technologies available for structural inspection in large diameter cast iron water mains was presented in Section 3.2. Eight technologies that have the potential to be used for structural inspection in large diameter cast iron pipes are described below. These technologies were used to evaluate the accuracy of the protocols by the authors based on their knowledge and the available reference material for the proposed technologies.

5.2   Tethered Remote Field Eddy Current

Remote field eddy current (RFEC) is a commonly used NDE testing method that can be used to measure pipe wall thickness in pipes made from various materials including cast iron (ASNT, 2004). RFEC technology uses an alternating current (AC) electromagnetic field generated by a coil that is concentric with the axis of the pipe. A sensor, or circumferentially distributed array of sensors, is placed near the inside of the pipe wall, displaced from the source of the AC field wall. The through-wall nature of the technique allows external and internal defects to be detected with approximately equal sensitivity. As the pipe wall thickness increases, the AC frequency must be reduced to maintain comparable defect sensitivity. However, when lower frequency signals are used, the inspection speed must be reduced accordingly.

Tools for boiler tube inspection are readily available for small pipes less than 16 in. in diameter. Systems for larger diameter pipe such as large cast iron greater than 16 in. are less common. For large diameter pipes, the tools have to be long since the field source and sensors have to be separated more than two pipe diameters.

The tool can be free swimming or tethered on a wire line. When wire line tethered, lengths up to a few thousand feet can be inspected from one launch point, limited by the number of bends and other factors that affect pull force. The free swimming version can inspect longer distances, and are often limited by obstructions and water system pipe configurations.

Protocol 1: Basic Screening
Protocol 1 is used to answer whether tethered RFEC is feasible for use in large diameter cast iron water mains. First, the suitability to large diameters and the intended capabilities are assessed as outlined in Protocol 1. A tool has recently been built and used for inspections in a 24 in. diameter pipeline (Nestleroth, et al., 2010). The tool is intended to detect wall thickness and thinning caused by corrosion or erosion, as well as joint couplings, branches, and elbows. Next, the primary category from Section 1.0 is selected based on its intended use. The tool is intended to be inserted into the pipe; therefore internal inspection (Section 1.1) is selected. The pipe does not have to be cleaned to base metal; the tool can operate in pipes with tuberculation that does not extend into the main more than 1 in. (Section 1.1.1), and the tool will be temporarily installed in the pipe (Section 1.1.2). Section 1.1.4 questions are answered for use in Protocol 2 as shown in the inline applicability grade card in Table 5-1. Based on the inline applicability grade card, the technology would be ***difficult to implement*** in water mains.

Table 5-1. Inline Applicability Grade Card for Tethered RFEC

Question	Rating	Comment
1.1.4.1	D	The system as proposed is pulled through the main
1.1.4.2	C	High accuracy RFEC needs many sensors, making the tool diameter > ½ the pipe diameter
1.1.4.3	C	Launch angle parallel to the pipe
1.1.4.4	C	Receive angle parallel to the pipe
1.1.4.5	C	Pipe must be cut
1.1.4.6	D	Pipe must be taken out of operation
1.1.4.7	A	Flow from branches not an issue
1.1.4.8	B	Large protrusions are of some concern
1.1.4.9	C	Butterfly valves and bends are a problem
1.1.4.10	C	Obstructions need to be known

Score: *Implementation factor Difficult: Mostly Bs and Cs (or worse)*

Protocol 2: Secondary Screening

Protocol 2 is used to determine if tethered RFEC can locate the key distress indicators for large diameter cast iron water mains as identified in Section 2. Tethered RFEC is used to detect wall thickness which is a form of degradation that could lead to failure (Section 2.1). The wall thickness could indicate corrosion in the pipe barrel (Section 2.1.1). The tool has the theoretical capability to find and quantify potentially service limiting wall loss (Section 2.1.1) and the accuracy depends on implementation factors and design compromises needed to adapt to pipeline inspection constraints. This remote field implementation can only examine the barrel of the pipe; hence Sections 2.1.2, 2.1.3, 2.2, and 2.3 are not applicable for this technology. Section 2.4 determines the cost grade for the technology as shown in Table 5-2. Based on the cost grade card, the technology has a *medium cost* to develop and implement in large diameter cast iron water mains.

Table 5-2. Cost Grade Card for Tethered RFEC

Question	Rating	Comment
2.4.1	B	Capital cost between $50,000 and $200,000
2.4.2	A	Less than $100,000 of additional developmental capital
2.4.3	C	Large tools mean more labor intensive
2.4.4	C	A dedicated truck is required
2.4.5	C	Detailed data requires a month to process
2.4.6	C	The inspection distance is the distance between valves, nominally 2,000 ft.
2.4.7	B	The cost to modify the pipeline is medium

Score: *Implementation cost Medium: Mostly Bs with a few As or Cs (or worse)*

Protocol 3: Tertiary Screening

Protocol 3 is used to determine if the tethered RFEC tool should or should not be developed further for use in large diameter cast iron water mains. Since the tool is designed to detect barrel defects from inside the main, Protocol 3 leads the user to Section 3.1.4. While the tool has the potential to detect and size wall loss in the barrel of the pipe, the method is difficult to implement and the cost to implement is medium. Therefore, tethered RFEC *is not considered appropriate for further development* for large diameter cast iron water mains. If this technology could inspect the bell for cracks or corrosion, the tethered RFEC may be considered appropriate for further development.

5.3   Robotic RFEC

Robotic RFEC is an untethered, remote-controlled robot for inspection of live natural gas mains. The tools have been designed to be used for visual inspection of cast iron and steel gas mains and can be equipped with an RFEC system to detect barrel corrosion by Carnegie Mellon's National Robotics

Engineering Center (NREC, 2011). The robot can be launched into the pipeline under live conditions and is designed to negotiate diameter changes, 45° and 90° bends, and tees. Like the tethered RFEC system, tool length can be an issue for larger diameter lines. One of the benefits of a robotic system is the inspection speed can be controlled, with higher speeds used for screening and lower speeds for detailed assessment.

Protocol 1: Basic Screening

Robotic RFEC is suitable for cast iron pipe, but it is currently used in gas mains up to 8 in. in diameter. The tool is intended to detect barrel corrosion while performing a visual inspection as well. The primary category from Section 1.0 is internal inspection (Section 1.1). The tool could work in pipes with tuberculation and sediment (Section 1.1.1), and the tool will be temporarily installed in the pipe (Section 1.1.2). Section 1.1.4 questions are answered for use in Protocol 2 as shown in the inline applicability grade card in Table 5-3. Based on the inline applicability grade card, the technology would be *moderate to implement* in water mains.

Table 5-3. Inline Applicability Grade Card for Robotic RFEC

Question	Rating	Comment
1.1.4.1	C	The system as proposed is a robotic crawler
1.1.4.2	B	The tool diameter is nominally equal to ½ the pipe diameter
1.1.4.3	A	Launch angle perpendicular to the pipe
1.1.4.4	A	Receive angle perpendicular to the pipe
1.1.4.5	A	Fitting can be installed while the pipe is pressurized
1.1.4.6	B	Line is full, but not operational
1.1.4.7	A	Flow from branches not an issue
1.1.4.8	B	Large protrusions are of some concern
1.1.4.9	C	Butterfly valves and bends may be a problem
1.1.4.10	A	Obstructions do not need to be known

Score: *Implementation factor Moderate: Mostly Bs with a few As or Cs (or worse)*

Protocol 2: Secondary Screening

Robotic RFEC is used to detect barrel corrosion, which is a form of degradation that could lead to failure (Section 2.1). The tool has the theoretical capability to find and quantify potentially service-limiting wall loss (2.1.1); the accuracy depends on implementation factors and design compromises needed to adapt to pipeline inspection constraints. Sections 2.1.2, 2.1.3, 2.2, and 2.3 are not applicable for this technology. Section 2.4 determines the cost grade for the technology as shown in Table 5-4. Based on the cost grade card, the technology has a *high cost* to develop and implement in large diameter cast iron water mains.

Table 5-4. Cost Grade Card for Robotic RFEC

Question	Rating	Comment
2.4.1	C	Capital cost between $200,000 and $500,000
2.4.2	B	Between $100,000 and $400,000 of additional developmental capital
2.4.3	C	Large tools mean more labor intensive
2.4.4	C	A dedicated truck is required
2.4.5	B	Onsite analysis available for identifying significant anomalies and detailed data requires a month to process
2.4.6	C	The inspection distance is the distance between valves, nominally 2,000 ft.
2.4.7	B	The cost to modify the pipeline is medium

Score: *Implementation cost High: Mostly Bs and Cs (or worse)*

Protocol 3: Tertiary Screening
Since robotic RFEC is designed to detect barrel corrosion from inside the main, Protocol 3 leads the user to Section 3.1.4. The tool can size wall loss anomalies, and since implementation is moderate despite the high cost to implement, ***the approach has the potential for further development***.

5.4   Free Swimming Acoustics

The free swimming acoustic (FSA) inspection system is an autonomous inline system that uses miniature electronic data acquisition systems and acoustic technology to detect and locate leaks and gas pockets in a pipeline and to assess average wall thickness. The FSA consists of two primary components: a core with data recording hardware and a lightweight shell for cushioning the device and propelling the unit in the pipe. The core houses the acoustic sensors, tracking equipment, data storage equipment, and power supply. The core is placed within the shell that can vary in diameter depending on the size, operation, and configuration of the pipeline to be surveyed. The shell is usually less than one third of the diameter of the pipe and can negotiate most pipeline obstructions such as valves as long as they are open.

Protocol 1: Basic Screening
Protocol 1 was used to determine if the FSA was feasible for use in large diameter cast iron water mains, which it is. The tool is intended to detect leaks and assess pipe wall thickness. The tool is intended to be inserted into the pipe; therefore internal inspection (Section 1.1) is selected. The pipe does not have to be cleaned to base metal; the tool can operate in pipes with tuberculation and sediment (Section 1.1.1), and the tool will be temporarily installed in the pipe (Section 1.1.2). Section 1.1.4 questions are answered for use in Protocol 2 as shown in the inline applicability grade card in Table 5-5. Based on the inline applicability, the technology would be ***moderate to implement*** in water mains.

Table 5-5. Inline Applicability Grade Card for FSA

Question	Rating	Comment
1.1.4.1	A	The system is free swimming, propelled by water flow
1.1.4.2	A	Small tool diameter, can be ≤ 6 in.
1.1.4.3	A	Launch angle perpendicular to the pipe
1.1.4.4	A	Receive angle perpendicular to the pipe
1.1.4.5	A	Fitting can be installed while the pipe is pressurized
1.1.4.6	A	Pipe can be full and operational
1.1.4.7	D	Flow from branches < ¼ of the main diameter must be stopped
1.1.4.8	A	Protrusions are of little concern
1.1.4.9	B	Butterfly valves must be fully opened
1.1.4.10	A	Obstructions do not need to be known

Score: *Implementation factor Moderate: Mostly Bs with a few As or Cs (or worse)*

Protocol 2: Secondary Screening
Protocol 2 is used to determine if the FSA can locate the key distress indicators for large diameter cast iron water mains as identified in Section 2. The FSA is used to detect leaks, which is a form of degradation that could lead to failure (Section 2.1). The tool does not detect whether or not the leak is at a joint, crack, or barrel of the pipe, but it has the potential to perform better than many existing methods (Section 2.1.3). Sections 2.2 and 2.3 are not applicable for this technology and Section 2.4 determines the cost grade for the technology as shown in Table 5-6. Based on the cost grade card, the technology has a ***medium cost*** to develop and implement in large diameter cast iron water mains.

Table 5-6. Cost Grade Card for FSA

Question	Rating	Comment
2.4.1	A	Capital cost less than $50,000
2.4.2	A	Less $100,000 of additional developmental capital
2.4.3	B	Low labor intensive (2 technicians)
2.4.4	B	Equipment can be shipped in a container
2.4.5	B	Detailed data requires two weeks to process
2.4.6	A	The inspection distance can be more than 10,000 ft
2.4.7	B	The cost to modify the pipeline is medium

Score: *Implementation cost Medium: Mostly Bs with a few As or Cs (or worse)*

Protocol 3: Tertiary Screening
Protocol 3 is used to determine if the FSA should be (or needs to be) developed further for use in large diameter cast iron water mains. Since the tool is designed to detect leaks from inside the main, Protocol 3 leads the user to Section 3.3.4. The tool locates leaks and has a cost grade less than high, so *the technology has the potential to be developed further*. The development should include the capability to detect whether a leak is a joint, crack, or in the barrel of the pipe.

5.5   Flexible Rod Sensor

Flexible rod based systems can be used to move sensors in the pipe both up and down stream of the insertion point. Using systems similar to those used to fish or snake wires in walls in the building industry, distances from the insertion point can be hundreds of feet. For these protocols, the flexible rod sensor (FRS) is a tethered system that can be used for leak detection and video assessment of water transmission mains. The system is able to determine the location of leaks and at the same time allows for the detection of tuberculation, liner condition, and service and valve placement.

Protocol 1: Basic Screening
The FRS is suitable for large diameter cast iron pipe, and can be used on pressurized trunk mains. The tool is intended to detect water leaks while performing a visual inspection as well. The primary category from Section 1.0 is internal inspection (Section 1.1). The tool could work in pipes with tuberculation and sediment (Section 1.1.1), and the tool will be temporarily installed in the pipe (Section 1.1.2). Section 1.1.4 questions are answered for use in Protocol 2 as shown in the inline applicability grade card in Table 5-7. Based on the inline applicability grade card, the technology would be *easy to implement* in water mains.

Table 5-7. Inline Applicability Grade Card for FRS

Question	Rating	Comment
1.1.4.1	B	The system is tethered
1.1.4.2	A	Small tool diameter, can be ≤ 6 in.
1.1.4.3	A	Launch angle perpendicular to the pipe or using hydrants
1.1.4.4	A	Receive angle perpendicular to the pipe or using hydrants
1.1.4.5	A	Fitting can be installed while the pipe is pressurized
1.1.4.6	A	Pipe can be full and operational
1.1.4.7	A	Flow from branches not an issue
1.1.4.8	A	Protrusions are of little concern
1.1.4.9	B	Butterfly valves must be fully opened
1.1.4.10	A	Obstructions do not need to be known

Score: *Implementation factor Easy: Mostly As and no Cs (or worse)*

Protocol 2: Secondary Screening
The FRS is used to detect leaks, which is a form of degradation that could lead to failure (Section 2.1). The tool may be able to identify if a leak is at a joint since it works with a CCTV camera as well (Section 2.1.3). Sections 2.2 and 2.3 are not applicable for this technology. Section 2.4 determines the cost grade for the technology as shown in Table 5-8. Based on the cost grade card, the technology has a *medium cost* to develop and implement in large diameter cast iron water mains.

Table 5-8. Cost Grade Card for FRS

Question	Rating	Comment
2.4.1	A	Capital cost less than $50,000
2.4.2	A	Less $100,000 of additional developmental capital
2.4.3	B	Low labor intensive (2 technicians)
2.4.4	B	Equipment can be shipped in a container
2.4.5	B	Detailed data requires two weeks to process
2.4.6	C	The inspection distance is the distance between valves, nominally 2,000 ft.
2.4.7	A	Potentially no or minimal cost to modify the pipeline

Score: *Implementation cost Medium: Mostly Bs with a few As or Cs (or worse)*

Protocol 3: Tertiary Screening
The FRS is designed to detect leaks from inside the main, so Protocol 3 leads the user to Section 3.3.4. The tool can measure leak rates and identify leak type, and since implementation is easy and cost to implement is medium, *the approach has the potential for further development*. The development should include verification of the system's ability to detect the location of the leak.

5.6   Magnetic Tomography

Magnetic tomography (MTM) is an emerging technology that makes magnetic measurements using sensitive magnetometers from above ground to assess the structural integrity of the pipeline. The method does not directly detect pipeline anomalies; rather it detects the increased level of stress caused by the internal pressure. While data on minimum detectable flaw size are not available, for older large diameter water mains that have wall thicknesses greater than a half inch and operating pressures less than 100 psi (which is very low pressure for a transmission main), the corrosion size would have to be substantial to be detected. Data are collected by a non-contact scanning magnetometer and are subsequently analyzed. The inspection record provides the location and extent of corrosion defects and other stress risers. The method works best on higher pressure transmission pipelines. Accuracy may be affected by either the close proximity of other pipelines and power lines, and would have to be investigated.

Protocol 1: Basic Screening
MTM may be suitable for large diameter cast iron pipe. The tool is intended to detect wall corrosion from the ground surface. The primary category from Section 1.0 is above ground inspection (Section 1.3). The tool does not require electrical conductivity of the pipe and technology works through pavement.

Protocol 2: Secondary Screening
MTM is used to find stress risers, which is a form of degradation that could lead to failure (Section 2.1). Significant corrosion in the barrel would be the most detectable anomaly, but stress in the bell due to misalignment may also be detectable (Section 2.1.1). Section 2.4 determines the cost grade for the technology as shown in Table 5-9. Based on the cost grade card, the technology has a *medium cost* to develop and implement in large diameter cast iron water mains.

Table 5-9. Cost Grade Card for MTM

Question	Rating	Comment
2.4.1	B	Capital cost between $50,000 and $200,000
2.4.2	B	Between $100,000 and $400,000 of additional developmental capital
2.4.3	B	Low labor intensive (2 technicians)
2.4.4	B	Equipment can be shipped in a container
2.4.5	C	Detailed data requires a month to process
2.4.6	D	The inspection process is slow, rate less than 1,000 ft per day
2.4.7	A	No cost to modify the pipeline

Score: *Implementation cost Medium: Mostly Bs with a few As or Cs (or worse)*

Protocol 3: Tertiary Screening

MTM is designed to detect primarily average wall thickness variation; therefore Protocol 3 leads the user to Section 3.1.1, and potentially bell corrosion (3.2.1) and effects of pipe angle (3.4.1). The tool locates these anomalies and has a cost grade less than high, so *the technology has the potential to be developed further* for large diameter cast iron water mains. One of the first steps in this development should be a sensitivity study to determine the size of anomalies that are detectable. This method is likely to screen pipes to determine which pipes should be excavated for detailed analysis.

5.7   Multi-Frequency Field Variation

Multi-frequency field variation (MFFV) was investigated for oil and gas pipelines to detect corrosion. It uses a current at high and low frequencies impressed onto the pipe over typically less than a few kilometers. An above ground magnetic field sensor array is used to detect field changes related to anomalies in the pipeline. It was tested by some pipeline transmission companies as a screening technique. It was reported to have merits, but not as a detailed pipeline integrity assessment method in the same way that inline inspection is used by that industry. Therefore, it was not further developed.

Protocol 1: Basic Screening

MFFV may be suitable for large diameter pipe. The tool is intended to detect field changes related to anomalies in the pipe from the ground surface. The primary category from Section 1.0 is above ground inspection (Section 1.3). The tool does require an above ground electrical connection of the pipe, and typically uses the cathodic protection systems used by oil and gas systems. Since these connections are not common on water systems, *this approach is not appropriate for further development* for use in large cast iron water mains.

5.8   MFL Inline Free Swimming Pig

Inline inspection is an integral part of many oil and gas pipeline company integrity management plans. The most common inspection technology for both natural gas and liquid pipelines is MFL. MFL was first used in the 1960s and was significantly improved in the 1980s and 1990s. While improvements are still being implemented, the performance capability of MFL tools has remained relatively unchanged for a decade. The major attribute of MFL is the ruggedness of the implementations that enable this technology to perform under the rigors presented by the pipeline environment. This technology can locate and size metal loss anomalies. The nominal depth sizing specification of most MFL inline tools is a tolerance of +/-10% of wall thickness with a certainty of 80% (4 of 5 depth readings are within the tolerance). The method can work through cement liners, but with degraded performance.

MFL tools are typically propelled through the pipeline by the product flow. Since water pipelines do not have simple methods for inserting the tool into the pipe and retrieving the tool, application of this method could be difficult. This technology has been offered by the oil and gas company Rosen Inspection as well as the water inspection service provider Pure.

Protocol 1: Basic Screening
MFL inline inspection may be suitable for large diameter cast iron pipe. The tool is intended to detect wall corrosion from inside the pipe. The primary category from Section 1.0 is internal inspection (Section 1.1). The pipe would need to be cleaned to the internal coating to be used (Section 1.1.1) and the tool will be temporarily installed in the pipe (Section 1.1.2). Section 1.1.4 questions are answered for use in Protocol 2 as shown in the inline applicability grade card in Table 5-10. Based on the inline applicability grade card, the technology would be *difficult to implement* in water mains.

Table 5-10. Inline Applicability Grade Card for Inline MFL

Question	Rating	Comment
1.1.4.1	A	The system is free swimming, propelled by water flow
1.1.4.2	D	The tool diameter is nominally the pipe diameter
1.1.4.3	C	Launch angle parallel to the pipe
1.1.4.4	C	Receive angle parallel to the pipe
1.1.4.5	C	Fitting cannot be installed while the pipe is pressurized
1.1.4.6	A	Pipe can be full and operational
1.1.4.7	B	Flow from branches between ½ and ¾ of the main diameter must be stopped
1.1.4.8	D	Any protrusion is of some concern
1.1.4.9	C	Butterfly valves and tight bends are a problem
1.1.4.10	C	Obstructions need to be known

Score: *Implementation factor Difficult: Mostly Bs and Cs (or worse)*

Protocol 2: Secondary Screening
MFL inline inspection could be used to detect barrel corrosion, which is a form of degradation that could lead to failure (Section 2.1). The tool can find and quantify potentially service limiting wall loss (2.1.1). Sections 2.1.2, 2.1.3, 2.2, and 2.3 are not applicable for this technology. Section 2.4 determines the cost grade for the technology as shown in Table 5-11. Based on the cost grade card, the technology has a *high cost* to develop and implement in large diameter cast iron water mains.

Table 5-11. Cost Grade Card for Inline MFL

Question	Rating	Comment
2.4.1	C	Capital cost between $200,000 and $500,000
2.4.2	A	Less than $100,000 of additional developmental capital
2.4.3	C	Large tools mean more labor intensive
2.4.4	C	A dedicated truck is required
2.4.5	B	Detailed data requires two weeks to process
2.4.6	C	The inspection distance is the distance between valves, nominally 2,000 ft
2.4.7	C	The cost to modify the pipeline is high

Score: *Implementation cost High: Mostly Bs and Cs (or worse)*

Protocol 3: Tertiary Screening
Since MFL inline inspection is designed to detect barrel corrosion from inside the main, Protocol 3 leads the user to Section 3.1.4. The tool can size wall loss anomalies, but since implementation is difficult and the cost to implement is high, *the approach is not appropriate for further development*.

5.9 Tethered MFL

While most MFL tools are typically propelled through the pipeline by the product flow, some have been designed to work with a pull cable or tether. Some tools are based on well casing inspection tools, while others are variations on free swimming pigs. Inspection companies have offered this as a service for municipal water and sewer lines as well as nuclear feed water lines with large diameters. These tools have a similar performance specification as free swimming MFL tools. Often, these tools have not been designed to pass tight bends and obstructions.

Protocol 1: Basic Screening

MFL inline pull-through inspection may be suitable for large diameter cast iron pipe. The tool is intended to detect wall corrosion from inside the pipe. The primary category from Section 1.0 is internal inspection (Section 1.1). The pipe would need to be cleaned to the internal coating to be used (Section 1.1.1) and the tool will be temporarily installed in the pipe (Section 1.1.2). Section 1.1.4 questions are answered for use in Protocol 2 as shown in the inline applicability grade card in Table 5-12. Based on the inline applicability grade card, the technology would be *difficult to implement* in water mains.

Table 5-12. Inline Applicability Grade Card for Tethered MFL

Question	Rating	Comment
1.1.4.1	B	The system as proposed will be pulled through the main
1.1.4.2	D	The tool diameter is nominally the pipe diameter
1.1.4.3	C	Launch angle parallel to the pipe
1.1.4.4	C	Receive angle parallel to the pipe
1.1.4.5	C	Fitting cannot be installed while the pipe is pressurized
1.1.4.6	B	Pipe can be full but not operational
1.1.4.7	A	Flow from branches not an issue
1.1.4.8	D	Any protrusion is of some concern
1.1.4.9	C	Butterfly valves and tight bends are a problem
1.1.4.10	C	Obstructions need to be known

Score: *Implementation factor Difficult: Mostly Bs and Cs (or worse)*

Protocol 2: Secondary Screening

MFL inline pull-through inspection could be used to detect barrel corrosion, which is a form of degradation that could lead to failure (Section 2.1). The tool can find and quantify potentially service limiting wall loss (Section 2.1.1). Sections 2.1.2, 2.1.3, 2.2, and 2.3 are not applicable for this technology. Section 2.4 determines the cost grade for the technology as shown in Table 5-13. Based on the cost grade card, the technology has a *high cost* to develop and implement in large diameter cast iron water mains.

Table 5-13. Cost Grade Card for Tethered MFL

Question	Rating	Comment
2.4.1	C	Capital cost between $200,000 and $500,000
2.4.2	A	Less than $100,000 of additional developmental capital
2.4.3	C	Large tools mean more labor intensive
2.4.4	C	A dedicated truck is required
2.4.5	B	Detailed data requires two weeks to process
2.4.6	C	The inspection distance is the distance between valves, nominally 2,000 ft
2.4.7	C	The cost to modify the pipeline is high

Score: *Implementation cost High: Mostly Bs and Cs (or worse)*

Protocol 3: Tertiary Screening
MFL inline pull-through inspection is designed to detect barrel corrosion from inside the main, so Protocol 3 leads the user to Section 3.1.4. The tool can size wall loss anomalies, but since implementation is difficult and the cost to implement is high, *the approach is not appropriate for further development*.

5.10   Application Summary

The application of the three protocols to the eight technologies above is intended to demonstrate the applicability of the protocols to various technologies. The two RFEC technologies had differing results since one was considered to be more difficult to implement. Both internal leak technologies were considered appropriate for further development, and both are currently being designed as tools for use on large cast iron water mains. The two above ground wall corrosion detection technologies had differing results as well. The first technology (MTM) was easily implementable on large diameter cast iron for a moderate cost, but the second technology (MFFV) required an above ground electrical connection of the pipe, which is not applicable on cast iron water mains. The final two technologies were both based on MFL technology and neither was considered to be appropriate for further development as both were determined to be difficult to implement for a high cost. A summary of the application for the eight technologies described above is shown in Table 5-14.

Table 5-14. Summary of Protocol Application

Technology	Intended Use	Applicability Grade	Cost Grade	Decision
Tethered RFEC	Barrel Corrosion	Difficult (1A,1B,6C,2D)	Medium (1A,2B,4C,0D)	Not appropriate for further development.
Robotic RFEC	Barrel Corrosion	Moderate (5A,3B,2C,0D)	High (0A,3B,4C,0D)	Has potential for further development.
Free Swimming Acoustic System	Leaks	Moderate (8A,1B,0C,1D)	Low (3A,4B,0C,0D)	Has potential for further development.
Flexible Rod Sensor	Leaks	Easy (8A,2B,0C,0D)	Medium (3A,3B,1C,0D)	Has potential for further development.
Magnetic Tomography	Barrel Corrosion	N/A	Medium (1A,4B,1C,1D)	Has potential for further development.
Multi-frequency field variation	Barrel Corrosion	N/A	N/A	Not appropriate for further development.
Free Swimming MFL	Barrel Corrosion	Difficult (2A,1B,5C,2D)	High (1A,1B,5C,0D)	Not appropriate for further development.
Tethered MFL	Barrel Corrosion	Difficult (1A,2B,5C,2D)	High (1A,1B,5C,0D)	Not appropriate for further development.

6.0: CONCLUSIONS AND RECOMMENDATIONS

6.1 Summary

High-risk, large diameter cast iron water mains are very costly when they fail, creating the need to determine how much longer these mains can safely operate. These mains are also expensive to replace and sound pipe that is replaced significantly before the end of its service life is a waste of limited resources. Structural inspection technologies are an important factor in determining the current and future condition of these water mains, although inspection can be expensive. Many factors affect the performance and value of structural inspection technologies, so organizations interested in supporting inspection technology improvement should attempt a thorough, systematic assessment of innovative structural inspection technology improvement options.

This report describes the most common failure modes that occur in large diameter cast iron mains and outlines the distress indicators that could alert a utility that failure may be imminent. Technologies currently available for detecting these indicators are briefly discussed and referenced as are organizations funding research to develop new and innovative ways to inspect water mains. The three protocols outlined in Section 4 are useful in screening new technologies potentially applicable to structural inspection by determining their feasibility for water mains; applicability in detecting large diameter cast iron water main distress indicators; and whether or not the technology should be developed further based on its applicability and comparison with existing technologies. The strength of the protocols is the objective process for selecting technologies for development. The weaknesses of the protocols include:

- The threshold for the rating criteria are estimates and may need to be adjusted when initially using the protocol
- The examples use theoretical systems that may be proposed for development. The decision to develop a system is based on the details of the system. The example may not reflect actual systems.
- The current protocol could be strengthened by providing addition discussion and information on the preparation and cleanup requirements for in-line inspection.

The protocols are demonstrated on eight technologies to validate the approach.

6.2 Recommendations

The authors recommend that EPA and other organizations interested in supporting the development of structural inspection technologies use the aforementioned protocols as a screening measure to determine if a proposed technology is applicable to large diameter cast iron mains, capable of detecting their key distress indicators, and implementable at a reasonable cost as to be potentially used by water utilities. This process was developed for large diameter cast iron mains as an example and can be expanded to small diameter mains and other pipe types. The authors recommend that screening protocols be developed for other potentially high risk mains such as large diameter ductile iron, prestressed concrete cylinder pipe (PCCP), asbestos cement, and steel.

7.0: REFERENCES

Al-Barqawi, H. and T. Zayed. 2006. "Condition Rating Model for Underground Infrastructure Sustainable Water Mains," *Journal of Performance of Constructed Facilities*, 20(2), 126-135.

American Society for Nondestructive Testing (ASNT). 2004. *Nondestructive Testing Handbook, Third Edition: Volume 5, Electromagnetic Testing*, Columbus, OH, ASNT.

American Water Works Association (AWWA). 2004. "Water:\\Stats 2002 Distribution Survey." AWWA, Denver, CO.

Baird, G. 2010. "A Game Plan for Aging Water Infrastructure." *Journal AWWA*, 102(4), 74-82.

Booth, G., A. Cooper, P. Cooper, and D. Wakerley. 1967. "Criteria of Soil Aggressiveness Towards Buried Metals: Experimental Methods." *British Corrosion Journal*, 2(3), 104-108.

Cassa, A.M. (2008). "A numerical investigation into the behavior of leak openings in pipes under pressure." M.I. Thesis, University of Johannesburg, South Africa.

Cast Iron Pipe Research Association (CIPRA). 1927. *Handbook of Cast Iron Pipe*. CIPRA, Chicago, IL.

Ductile Iron Pipe Research Association (DIPRA). 2005. *The Design Decision Model for Corrosion Control of Ductile Iron Pipelines*. DIPRA.

Ferguson, P. and D. Downey. 2009. "Soil Testing for Condition Assessment of Buried Mains." *Trenchless Technologies in Asia Pacific*, CHKSTT, Hong Kong.

Ferguson, P. and D. Nicholas. 1984. "Accurate Prediction of Cast Iron Water Main Performance."

Glaser, S. and D. Pescovitz. 2002. "National Workshop on Future Sensing Systems: Living, Nonliving, and Energy Systems." NSF Workshop, Washington, D.C.

Hannaford, M., W. Melia, P. Hoyt, and R. Jackson. 2010. "An Advanced Method of Condition Assessment for Large-Diameter Mortar-Lined Steel Pipelines." *AWWA ACE*, Chicago, IL.

Jarvis, M. and M. Hedges. 1994. "Use of Soil Maps to Predict the Incidence of Corrosion and the Need for Iron Mains Renewal." *Water and Environmental Journal*, 8(1), 68-75.

Jason Consultants. 2007. *Inspection Guidelines for Ferrous Force Mains*. 04-CTS-6UR, WERF, Alexandria, VA.

Kleiner, Y. and B. Rajani. 2000. "Considering Time-dependent Factors in the Statistical Prediction of Water Main Breaks." *AWWA Infrastructure Conference*, Baltimore, MD, pp. 1-12.

Kleiner, Y., B. Rajani, and R. Sadiq. 2005. "Risk Management of Large Diameter Transmission Water Mains." AWWARF, Denver, CO.

Kundu, T. 2005. "Development of Non-contact Sensors for Pipe Inspection by Lamb Waves." Project No. 9901221, NSF, Washington, D.C.

Lillie, K., C. Reed, M. Rodgers, S. Daniels, and D. Smart. 2004. "Workshop on Condition Assessment Inspection Devices for Water Transmission Mains." AWWARF Project No. 2871, Denver, CO.

Makar J. 1999a. "Failure Analysis for Grey Cast Iron Water Pipes." *Distribution System Symposium*, Reno, NV, Sept. 19-21, AWWA.

Makar J. 1999b. "IRC Researchers Investigate Cast Iron Pipe Failures." *Construction Innovation*, 4(3), NRC Canada.

Makar J. 2000. "Prone to Fail." *Canadian Consulting Engineer*, 41(7), 56-58.

Makar, J. 2001. "Investigating Large Gray Cast Iron Pipe Failures: A Step by Step Approach." *AWWA Infrastructure*, Orlando, FL.

Makar, J., R. Desnoyers, and S. McDonald. 2001. "Failure Modes and Mechanism in Gray Cast Iron Pipes." *Underground Infrastructure Research*, Waterloo, ON.

Makar, J. 2005. "The Effect of Corrosion Pitting on Circumferential Failures in Gray Cast Iron Pipes." AWWARF, Denver, CO.

Marshall, P. 2000. "Understanding Burst Rate Patterns of Water Pipes." UKWIR, London, UK.

Marshall, P. 2001. "The Residual Structural Properties of Cast Iron Pipes: Structural and Design Criteria for Lining for Water Mains in Pipeline Innovation." UKWIR, London, UK.

Moser A. 2008. *Buried Pipe Design*. McGraw Hill.

Mu, J. 2011. "In-situ Imaging of Water Pipelines using Ultrasonic Guided Waves." EPA Small Business Innovation Research (SBIR), EPA Contract Number: EPD11041.

National Institute of Standards and Technology (NIST). 2011. "Infrastructure Defect Recognition, Visualization, and Failure Prediction System Utilizing Ultra Wide Band Pulsed Radar Profilometry." TIP project, NIST.

National Robotics Engineering Center (NREC). 2011. "Pipeline Explorer." Accessed on August 12, Carnegie Mellon University.

Nestleroth, J. B., S. A. Flamberg, L. Wang, A. Chan, M. D. Royer, and A. F. Williams. 2010. "Field Demonstration of Emerging Pipe Wall Integrity Assessment Technologies for Large Cast Iron Water Mains." (Paper) In: Proceedings, American Society of Civil Engineers (ASCE) Pipelines 2010, Keystone, CO, August 28 - September 01, 2010. ASCE, Reston, VA, Paper 101.

Rajani, B., J. Makar, S. McDonald, C. Zhan, S. Kuraoka, C. Jen, and M. Viens. 2000. "Investigation of Grey Cast Iron Water Mains to Develop a Methodology for Estimating Service Life." AWWARF Project No. 280, Denver, CO.

Rajani, B. and Y. Kleiner. 2004. "Non-destructive Inspection Techniques to Determine Structural Distress Indicators in Water Mains." *Evaluation and Control of Water Loss in Urban Water Networks*, Valencia, Spain, pp. 1-20.

Rajani, B. and Y. Kleiner. In Press 2013. "Fracture Failure of Large Diameter Cast Iron Water Mains." WaterRF Project No. 4035, Denver, CO.

Reed, C., A. Robinson, and D. Smart. 2004. "Techniques for Monitoring Structural Behavior of Pipeline Systems." AWWARF Project No. 2612, Denver, CO.

Reed, C., A. Robinson, and D. Smart. 2006. "Potential Techniques for the Assessment of Joints in Water Distribution Pipelines." AWWARF Project No. 2689, Denver, CO.

Royer, M. 2005. *White Paper on Improvement of Structural Integrity Monitoring for Drinking Water Mains*. Office of Research and Development, Cincinnati OH. EPA/600/R-05/038.

Sears, E. (1964). "Ductile-iron pipe design." *Journal AWWA*, 56(1), 4-22.

Seica, M., J. Packer, M. Grabinsky, and B. Adams. 2002. "Evaluation of the Properties of Toronto Iron Water Mains and Surrounding Soil," *Can. J. of Civil Eng.*, 29(2), 222-237.

Stanton Ironworks. 1936. "Cast Iron Pipe: Its Life and Service." Stanton Ironworks Co., Nottingham, UK.

UK Water Industry Research (UKWIR). 2011. *A Survey of Practices for the Detection and Location of Leaks*. UKWIR 11/WM/08/45.

U.S. Environmental Protection Agency. 2001. *EPA Requirements for Quality Assurance Project Plans (QA/R-5)*. EPA/240/B-01/003, Office of Environmental Information, Washington, D.C. 40 pp.

U.S. Environmental Protection Agency. 2009. *Condition Assessment of Ferrous Water Transmission and Distribution Systems*. EPA/600/R-09/055. Office of Research and Development. Cincinnati, OH, 111 pp.

U.S. Environmental Protection Agency. 2010. *Control and Mitigation of Drinking Water Losses in Distribution Systems*. EPA/816/R-10/019, Office of Water, Washington, D.C. 176 pp.

U.S. Environmental Protection Agency. 2012a. *Condition Assessment Technologies for Water Transmission and Distribution Systems.* EPA/600/R-12/017. Office of Research and Development, Cincinnati, OH. 149 pp.

U.S. Environmental Protection Agency. 2012b. *Field Demonstration of Innovative Condition Assessment Technologies for Water Mains: Leak Detection and Location.* EPA/600/R-12/018, U.S. EPA, Office of Research and Development, Cincinnati, OH. 184 pp.

Water Environment Research Foundation (WERF). 2004. *An Examination of Innovative Methods Used in the Inspection of Wastewater Systems*. 01-CTS-7, WERF, Alexandria, VA.

Water Environment Research Foundation (WERF). 2007. *Condition Assessment Strategies and Protocols for Water and Wastewater Utility Assets*. 03-CTS-20CO, WERF, Alexandria, VA.

Water Research Foundation (WaterRF). 2011a. *Practical tool for deciding rehabilitation vs. replacement of cast iron pipes*. Project No. 4234, WaterRF, Denver, CO.

Water Research Foundation (WaterRF). 2011b. *Acoustic signal processing for pipe condition assessment*. Project No. 4360, WaterRF, Denver, CO.

APPENDIX A

ORGANIZATIONS FUNDING STRUCTURAL INSPECTION RESEARCH

ORGANIZATIONS FUNDING STRUCTURAL INSPECTION RESEARCH

A.1   U.S. Environmental Protection Agency.

Improving structural inspection technology capability through research and development (R&D), testing, and verification is a proactive, flexible approach to accomplishing a number of EPA's short- and long-term drinking water protection goals. Reducing high risk main breaks supports the Safe Drinking Water Act (SDWA) goals of protecting public health and drinking water quality. Reducing main breaks, optimizing maintenance planning, extending infrastructure service lives, and reducing water leakage support the goals of EPA for reducing the infrastructure funding gap and improving utilities' infrastructure management capability. The EPA provides avenues for structural inspection technology R&D through the: National Risk Management Research Laboratory's (NRMRL's) Water Supply and Water Resources Division (WSWRD); Environmental Technology Program (ETV); Small Business Innovation Research (SBIR); Center for Environmental Industry & Technology (CEIT); Clean Water Act (CWA); International Science and Technology Center (ITSC); and Science to Achieve Results (STAR) grants from the Office of Research and Development (ORD) National Center for Environmental Research (NCER).

A.2   U.S. Department of Transportation.

Under the direction of the DOT, the Office of Pipeline Safety (OPS) is the primary authority regarding the safety of natural gas and hazardous liquid pipelines for the large amount of energy product that is transported throughout the nation. The mission of OPS is to ensure that the operation of the nation's pipeline system is safe, reliable, and environmentally sound. The OPS conducts and supports research to maintain conformity with regulatory guidelines and provides tools and information regarding maintenance to maximize the impact on pipeline safety. The research and development projects focus on technologies for leak detection, improved system controls, prevention of damage, improvement of pipe materials, and monitoring.

A.3   U.S. Department of Energy.

The DOE provides almost 40% of total Federal funding in the area of research for energy, biological, computational, and environmental science. Most of the research is conducted by a variety of national laboratories, such as the National Energy Technology Laboratory (NETL) and technology centers. The Office of Scientific and Technical Information (OSTI) provides a searchable database that can be used to review ongoing or completed research and a search revealed that DOE has funded some projects related to structural inspection and leak detection in natural gas pipelines.

A.4   U.S. Department of Defense.

DoD has an in-house Research, Development, Test & Evaluation (RDT&E) program that applies basic research and advanced development of innovative technologies. There is also the Strategic Environmental Research and Development Program (SERDP), the corporate environmental research and development program executed in full partnership with the DOE and the EPA. Additionally, the Environmental Security Technology Certification Program (ESTCP) will demonstrate and validate promising technologies that target the DoD's most urgent environmental needs through implementation and commercialization. The result is a return on investment through savings in costs and improvement in efficiency. Successful demonstration of such technologies helps gain acceptance from regulatory communities and end-users. Additional avenues of research are through DoD laboratories and centers, including the: U.S. Army Corps of Engineers (USACE) Construction Engineering Research Laboratory

(CERL); Army Research Laboratory (ARL); and Nondestructive Testing, Information, and Analysis Center (NTIAC).

A.5   U.S. Department of Commerce.

The National Institute of Standards and Technology (NIST) is a non-regulatory federal agency within the DOC's Technology Administration. The mission of NIST is to develop and promote measurement, standards, and technology to enhance productivity, facilitate trade, and improve the quality of life. The two main cooperative programs used by NIST to meet their mission is the NIST Laboratories, which conducts research to advance the nation's technology infrastructure, and the Technology Innovation Program (TIP), which supports innovation through high-risk high-reward research in areas of critical need. One critical area identified as part of TIP is civil infrastructure, which includes new technologies for infrastructure inspection.

A.6   U.S. Department of Homeland Security.

Under the DHS, the Science and Technology Directorate (S&T) is the primary research and development agency for providing leading technologies to federal, state, and local officials for the protection of people and infrastructure from possible threats. DHS also uses the established National and Federal Laboratory system for development and research currently used by the DOT and DOE.

A.7   U.S. Department of the Interior.

The DOI's Bureau of Ocean Energy Management, Regulation, and Enforcement (BOEMRE) is the federal agency responsible for overseeing the safe and environmentally responsible development of energy and mineral resources on the Outer Continental Shelf. Under BOEMRE, the Technology Assessment & Research (TA&R) Program supports research associated with operational safety and pollution prevention, as well as oil spill response and cleanup capabilities. The TA&R program operates through contracts with universities, private firms, and government laboratories to assess safety-related technologies and to perform necessary applied research. A search revealed that one TA&R report addressed pipeline assessment methods relating to welded steel pipelines.

A.8   National Science Foundation.

NSF is an independent federal agency created by Congress in 1950. The purpose of NSF is to initiate and support scientific and engineering research through grants and contracts conducted at colleges and universities. NSF acts as a central agency for the collection, interpretation and analysis for all levels of scientific research. This information is provided to federal agencies for assistance in the generation of policies and procedures. NSF sponsors a broad range of research in the areas of NDE, sensors, materials, and other relevant topics. A search for applicable structural inspection systems and components conducted using the NSF award database for those dealing with pipe and NDE inspection techniques revealed a couple projects that could be applicable to large diameter cast iron water mains.

A.9   National Aeronautics and Space Administration.

NASA has a NDE Working Group that uses the Langley Research Center (LaRC) and the Ames Research Center for most of its research and development of NDE technologies. LaRC leads the major thrust of the NDE research program. The program focuses on maintaining an NDE science base core, developing new technologies for NASA, and transferring problem solutions to their clients. LaRC interacts with scientists, engineers, field centers, aerospace contractors, US industry, and universities. The LaRC NDE research program is focused in two offices (Safety and Mission Quality and Aero-Space Transportation Technology) which cover applications primarily for Space Operations/Transportation

System (spacecraft integrity), Subsonic, Supersonic and Hypersonic Aeronautics (aircraft integrity). There was some limited information regarding previous demonstrations, and testing of NDE technologies that theoretically could be applied to fatigue cracking and monitoring of ferromagnetic pipes.

A.10   Water Research Foundation.

WaterRF, formally known as AwwaRF, is a member-supported, international, nonprofit organization that sponsors research to enable water utilities, public health agencies, and other professionals to provide safe and affordable drinking water to consumers. WaterRF sponsors a scientific research program that is responsive to the needs of the water community by promoting the benefits of research and sharing the results with the community. WaterRF has a close partnership with EPA and has worked on nearly 200 projects. Several research projects have dealt with the reliability of cast iron water mains and the causes of deterioration, which have been consulted in preparation of this report.

A.11   Water Environment Research Foundation.

WERF is a non-profit organization that funds the development of independent scientific research dedicated to wastewater and stormwater issues. WERF operates with funding from subscribers and the federal government. Subscribers include wastewater treatment plants, stormwater utilities, and regulatory agencies. Industry, equipment companies, engineers, and environmental consultants also lend their support and expertise as subscribers. WERF takes a progressive approach to research, stressing collaboration among teams of subscribers, environmental professionals, scientists, and staff. All research is peer-reviewed by leading experts. The majority of the research programs fall under the broad categories of collection and treatment, infrastructure management, watersheds, ecosystems, and human health. The WERF program relevant to this report is the Strategic Asset Management (SAM) Challenge. The SAM Challenge seeks to evaluate and improve decision-making tools, techniques and methods to assist utilities in implementing asset management. Research under this challenge includes projects that have examined possible technologies for inspection and monitoring that could apply to large diameter cast iron water mains such as research into force main inspection.

A.12   Gas Technology Institute.

GTI is a not-for-profit research and development organization that funds the development and deployment of energy technology. GTI addresses key issues impacting natural gas and energy markets in the areas of energy supply, delivery, and end use and provides programs and services to industry, government, and consortia that include contract and collaborative R&D, technical services, and education programs. One key area of natural gas delivery research is pipeline integrity management, which includes technology research that could be applicable to large diameter cats iron main research. Included in this focus area are technologies used for: external and internal corrosion detection, inline inspection, and pipeline NDE such as broadband electromagnetic technology.

A.13   Industrial Research.

Several private companies and technology vendors continuously invest in the in the development of new technologies and improvement of existing technologies. For cast iron water mains, a recent example includes the Pressure Pipe Inspection Company (PPIC) PipeDiver®, which uses RFEC technology to generate magnetic currents in ferrous pipes for detection of pipe anomalies (EPA, 20011b). For the same EPA demonstration project, Russell NDE custom developed a 24-in. See Snake® RFT tool for measuring pipe wall thickness (Nestleroth, et al., 2010). Other technologies demonstrated as part of the EPA study included: PPIC's Sahara®; Pure's SmartBall™; Echologics' LeakfinderRT and ThicknessfinderRT;

Advanced Engineering Solutions, Ltd. External Condition Assessment Tool; and Rock Solid Group's Hand Scanning Kit and Crown Assessment Probe.

www.ingramcontent.com/pod-product-compliance
Lightning Source LLC
Chambersburg PA
CBHW081552170526
45166CB00009B/2675